Glencoe Science

BIOLOGY

The Dynamics of Life

Laboratory Manual
Student Edition

 **Glencoe**

New York, New York   Columbus, Ohio   Chicago, Illinois   Peoria, Illinois   Woodland Hills, California

A GLENCOE PROGRAM
BIOLOGY: THE DYNAMICS OF LIFE

Glencoe Online SCIENCE

Visit the Glencoe Science Web site
bdol.glencoe.com

You'll find:
Standardized Test Practice, Interactive
Tutor, Section and Chapter Self-Check
Quizzes, Online Student Edition, Web
Links, Microscopy Links, WebQuest
Projects, Internet BioLabs, In the News,
Textbook Updates, Teacher Bulletin
Board, Teaching Today

and much more!

Glencoe

Send all inquiries to:
Glencoe/McGraw-Hill
8787 Orion Place
Columbus, OH 43240-4027

ISBN 0-07-860228-9

Printed in the United States of America.

2 3 4 5 6 7 8 9 10   009   08 07 06 05 04

Contents

Chapter 1 Biology: The Study of Life

Chapter 2 Principles of Ecology

Chapter 3 Communities and Biomes

Chapter 4 Population Biology

Chapter 5 Biological Diversity and Conservation

Chapter 6 The Chemistry of Life

Chapter 7 A View of the Cell

Chapter 8 Cellular Transport and the Cell Cycle

Contents continued

Contents continued

How to Use This Laboratory Manual

Working in the laboratory throughout the course of the year can be an enjoyable part of your biology experience. **Biology: The Dynamics of Life,** *Laboratory Manual* is a tool for making your laboratory work both worthwhile and fun. The laboratory activities are designed to fulfill the following purposes:

- to stimulate your interest in science in general and especially in biology
- to reinforce important concepts studied in your textbook
- to allow you to verify some of the scientific information learned during your biology course
- to allow you to discover for yourself biological concepts and ideas not necessarily covered in class or in the textbook readings
- to acquaint you with a variety of modern tools and techniques used by today's biological scientists

Most importantly, the laboratory activities will give you firsthand experience in how a scientist works.

The activities in this manual are of two types: Investigation or Exploration. In an Investigation activity, you will be presented with a problem. Then, through use of controlled scientific methods, you will seek answers. Your conclusions will be based on your experimental observations alone or on those made by the entire class, recorded data, and your interpretation of what the data and observations mean. In the Exploration, you will make observations but will use fewer scientific methods to reach conclusions.

The general format of the activities in **Biology: The Dynamics of Life,** *Laboratory Manual* is listed below. Understanding the purpose of each of these parts will help make your laboratory experiences easier.

1. **Introduction**—The introductory paragraphs give you background information needed to understand the activity.

2. **Objectives**—The list of objectives is a guide to what will be done in the activity and what will be expected of you.

3. **Materials**—The materials section lists the supplies you will need to complete the activity.

4. **Procedure**—The procedure gives you step-by-step instructions for carrying out the activity. Many steps have safety precautions. Be sure to read these statements and obey them for your own and your classmates' protection. Unless told to do otherwise, you are expected to complete all parts of each assigned activity. Important information needed for the procedure but that is not an actual procedural step is also found in this section.

5. **Hypothesis**—In Investigations, you will write a hypothesis statement to express your expectations of the results and as an answer to the problem statement. Explorations do not require you to make a hypothesis.

6. **Data and Observations**—This section includes tables and space to record data and observations.

7. **Analysis**—In this section, you draw conclusions about the results of the activity just completed. Rereading the introduction before answering the questions is most helpful at this time.

8. **Checking Your Hypothesis**—You will determine whether your data supports your hypothesis.

9. **Further Investigations/Further Explorations**—This section gives ideas for further activities that you may do on your own. They may be either laboratory or library research.

Some of the Investigations are called Design Your Own, which are similar to the Design Your Own labs in your textbook. In Design Your Own Investigations, you will design your own experiments to find answers to problems. As with other Investigations, a Design Your Own Investigation includes introductory information and objectives. A *Problem* section presents the challenge posed by the Investigation. The *Hypotheses* section provides space for you to write your hypothesis, which you will test by using a procedure that you develop. The Possible Materials list includes items you could use in your procedure. Guidance in developing your procedure is provided in the *Plan the Experiment* section, and suggestions for checking your procedure are included in *Check the Plan*. A Design Your Own Investigation also includes *Data and Observations, Analysis, Checking Your Hypothesis,* and *Further Investigations* sections.

In addition to the activities, this laboratory manual has several other features—a glossary, a description of how to write a lab report, a section on the care of living things, diagrams of laboratory equipment, and information on safety that includes first aid and a safety contract. The glossary, included in the back of the manual, defines terms used throughout the manual. A pronunciation key has also been included to help you with the more difficult words. Read the section on safety now. Safety in the laboratory is your responsibility. Working in the laboratory can be a safe and fun learning experience. By using **Biology: The Dynamics of Life,** *Laboratory Manual*, you will find biology both understandable and exciting. Have a good year!

Writing a Laboratory Report

When scientists perform experiments, they make observations, collect and analyze data, and formulate generalizations about the data. When you work in the laboratory, you should record all your data in a laboratory report. An analysis of data is easier if all data are recorded in an organized, logical manner. Tables and graphs are often used for this purpose.

A written laboratory report should include all of the following elements.

TITLE: The title should clearly describe the topic of the report.

HYPOTHESIS: Write a statement to express your expectations of the results and as an answer to the problem statement.

MATERIALS: List all laboratory equipment and other materials needed to perform the experiment.

PROCEDURE: Describe each step of the procedure so that someone else could perform the experiment following your directions.

RESULTS: Include in your report all data, tables, graphs, and sketches used to arrive at your conclusions.

CONCLUSIONS: Record your conclusions in a paragraph at the end of your report. Your conclusions should be an analysis of your collected data.

Read the following description of an experiment, then answer the questions.

All plants need water, minerals, carbon dioxide, sunlight, and living space. If these needs are not met, plants cannot grow properly. A biologist thought that plants would not grow well if too many were planted in a limited area. To test this idea, the biologist set up an experiment. Three containers were filled with equal amounts of potting soil. One bean seed was planted in Container 1, five seeds in Container 2, and ten seeds in Container 3. All three containers were placed in a well-lit room. Each container received the same amount of water every day for two weeks. The biologist measured the heights of the growing plants every day. Then the average height of the plants in each container each day was calculated and recorded in a table. The biologist then plotted the data on a graph.

1. What was the purpose of this experiment?

_____

2. What materials were needed for this experiment?

_____

3. Write a step-by-step procedure for this experiment.

_____

_____

_____

_____

_____

_____

_____

_____

_____

_____

_____

4. Table 1 shows the data collected in this experiment. Based on these data, state a conclusion for this experiment.

_____

_____

_____

Table 1

Average Height of Growing Plants (in millimeters)										
	Day									
Container	1	2	3	4	5	6	7	8	9	10
1	20	50	58	60	75	80	85	90	110	120
2	16	30	41	50	58	70	75	80	100	108
3	10	12	20	24	30	35	42	50	58	60

5. Plot the data in Table 1 on a graph. Show average height on the vertical axis and the days on the horizontal axis. Use a different colored pencil for the graph of each container.

Care of Living Things

Caring for living things in a biology laboratory can be interesting and fun, and it can help develop the respect for all life that comes only from firsthand experience. In a room with an aquarium, terrarium, healthy animals, or growing plants, there is always some observable interaction between organisms and their environment. There are many species of plants and animals that are suitable for a classroom, but having them should be considered only if proper care will be taken so that the organisms not only survive, but thrive. Before growing plants or bringing animals into a classroom, find out if there are any health or safety regulations restricting their use or if there are any applicable state or local laws governing live plants and animals. Also, be sure not to consider cultivating any endangered or poisonous species. A biological supply house or local pet store will provide growing tips for plants or literature on animal care when these organisms are purchased.

Evaluating Resources

Before bringing any live specimens into a new environment, check with your teacher to see if their basic needs will be met in their new location. Plants need either sunlight or grow lights. Animals must be placed in well-ventilated areas out of direct sunlight and away from the draft of open windows, radiators, and air conditioners. For both animals and plants, a source of fresh water is essential. Consider what the fluctuation in temperature is over weekends and holidays and who will care for the plants or animals during those times.

Setting Up an Aquarium

A closed system such as an aquarium provides a variety of animals and plants and can be maintained easily if set up correctly. A 10- or 20-gallon tank can be a suitable home for about 5 to 10 tropical fish or even more of the temperate goldfish. An air pump, filter, heater, thermometer, and aquarium light (optional) need to be in working order. First fill an aquarium with a layer of gravel, then fill with water. If using tap water, let the water stand a day before putting any fish in the tank. During cooler months, adjust the thermostat of the heater to bring the water to the desired temperature before adding fish. Most fish require temperatures of 20° to 25°C. An inexpensive pH kit purchased from a pet store will test the acidity of the water and guide the maintenance of a healthy pH.

Choose fish that are compatible with one another. A pet store clerk can help in the selection. It is worth purchasing a scavenger fish, such as a catfish, or an algae eater that will help keep the tank clean of algae.

Snails are also helpful for this purpose. After purchasing, keep fish in the plastic bag containing water in which they came. Float the bag in the aquarium until the water reaches the same temperature, then slowly let the fish swim out of the bag. Some fish, such as guppies, eat their young. A smaller brood tank can be placed inside the aquarium to keep the mother separated from the young.

One person should be responsible for feeding the fish. Feed fish sparingly. Overfeeding is not healthy for the fish; also, it clouds the tank and causes unnecessary decay. Weekend or vacation food should also be available. These are slow-dissolving tablets that can feed the fish over vacations.

Plants can be added to an aquarium as well. *Elodea, Anacharia, Sagittaria, Cabomba,* or *Vallianeria* grown in a fish tank also are useful for many biology lab activities. Monitor their growth carefully and trim plants if growth is excessive. Some fish and snails may nibble on the plants, causing them to break apart and decay. Decay introduces bacterial populations that may endanger the fish, so be sure to remove any decaying plant matter.

Variations on an aquarium include setting up a "balanced aquarium" with fish, plants, and scavengers in balance so that no pump or filter is necessary. This usually takes more planning and maintenance than a filtered tank. More maintenance is needed also for a marine aquarium because of the corrosive nature of salt water. However, if specimens of marine organisms are readily available, creating such a mini-habitat is well worth the effort.

Setting Up a Terrarium

A terrarium is a mini-ecosystem that makes a suitable habitat for small plants and cold-blooded animals such as amphibians and reptiles. Start with a layer of gravel in the bottom of a 10-gallon tank. Add a layer of sand, then some topsoil. Plant small mosses and ferns in the topsoil and water moderately. At one end of the tank, bury a dish of water in the sand. This can be a water source for salamanders, frogs, or toads. Cover the terrarium. Check the humidity level regularly. If water droplets form inside the glass, or if water collects in the gravel, take the cover off for a while. Too much moisture will foster growth of mold. A specialized terrarium, such as a desert or bog, can be made by planting suitable plants and maintaining water conditions similar to those environments.

Animals such as newts and turtles require a terrarium with more water. For these types of animals, fill an aquarium one-third of the way with water and place rocks in it so that the animals can get out of the water. Other animals such as lizards, anoles, or chameleons can do well in a dry terrarium with rocks as long as they have a water bowl. Care should be taken to give cold-blooded animals enough heat during the cooler months. This can be done with a lamp. Feed cold-blooded animals according to their needs. Many enjoy live worms or insects, such as grasshoppers or crickets. However, freeze-dried worms available from a pet store are a good source of protein and require less maintenance than live food.

Keeping Mammals in a Classroom

Keeping mammals takes more consideration and commitment of time and expense. A small mammal such as a gerbil, guinea pig, hamster, or rabbit may be kept in a classroom, at least for a short time. Explore the possibility of dwarf breeds that are more at home in a small space. However, many mammals are sensitive, social mammals that form bonds and attachments to people. Life in a small cage alone most of the time is not suitable for a long and healthy life. For short periods of time, however, small animals may be kept in a cage, provided it is clean and large enough. Find out the exact nutritional needs of the animals; feed them on a regular schedule and provide fresh water daily. Some animals require dry food supplemented with fresh foods, such as greens. However, these foods spoil more rapidly and thus uneaten portions must be removed. Provide a large enough cage for the animal as well as materials for bedding, nesting, and gnawing. Clean the cages frequently. Letting urine and feces collect in a cage fosters the growth of harmful bacteria. Animals in a cage also require an exercise wheel. Lack of space combined with overeating can make an animal overweight and lethargic. Handle animals gently. Under no circumstances should animals be exposed to harmful radiation, drugs, toxic chemicals, or surgical procedures.

Many times students wish to bring a pet or even a wild animal that they have found into the classroom for observation. Do so only with discretion and if a proper cage is available. Protective gloves and glasses should be worn while handling any animals with the potential to bite. Be sure to check with local park rangers or wildlife specialists for any wildlife restrictions that may apply. Return any wild animals to their environment as soon as possible after observations.

Growing Plants in the Classroom

To successfully grow plants in a classroom, have on hand commercial potting soil, suitable containers such as clay or plastic pots, plant fertilizer, a watering can, and a spray bottle for misting. Always put a plant in the correct size container. One that is too large will encourage root growth at the expense of the stem and leaves. Place bits of broken clay or gravel in the bottom of the pot for drainage, then add potting soil and the plant. Place in a warm, well-lighted area and supply water. Give careful attention to a new plant to assess its adaptation to its new environment. Pale leaves may indicate insufficient light, yellowing leaves indicate overwatering, and dropping leaves usually indicate insufficient humidity. Fertilize only as directed.

With very little special attention, plants such as geraniums, begonias, and coleus can be easily and inexpensively grown in a classroom. These plants are hardy and can withstand fluctuations in light and temperature. From one hardy plant, many cuttings can be made to demonstrate vegetative propagation. A cutting of only a few leaves on a stem will develop roots in 2 to 4 weeks if it is placed in water or given root-growth hormone powder.

These plants not only add color to a classroom but are useful in biology experiments as well. The dense green leaves of geraniums are especially useful for extracting chlorophyll or showing the effects of light deprivation. The white portions of variegated coleus leaves are good for showing the absence of photosynthesis with a negative starch-iodine reaction. Pinch back the flower buds as they begin to form to encourage fuller leaf growth.

Larger plants such as a fig (*Ficus*), dumbcane (*Dieffenbachia*), cornplant (*Dracaena*), Norfolk Island pine (*Araucaria*), umbrella plant (*Schefflera*), or various philodendrons adapt well to low-light conditions and so do not need frequent watering. However, make sure humidity is suitable to avoid dropping leaves. More exotic plants might be best suited to a small-dish garden but will need special care because there is less soil to hold moisture.

During winter months, a dish garden of forced bulbs, such as paperwhite narcissus, can be easily grown by placing the bulbs in a container of water left in a cool, dark place. Blooms will appear in 3 to 4 weeks. In the early spring, shoots of early flowering shrubs, such as forsythia and pussy willow, may be forced. Cut off some healthy shoots when buds appear, wrap in wet newspaper, then bring indoors and immerse cut ends in a tall vase or jar. Also buds of fruit trees, such as apple, plum, or peach, will produce leaves and flowers in this way. Be sure to maintain shoots by changing water when necessary.

Laboratory Equipment

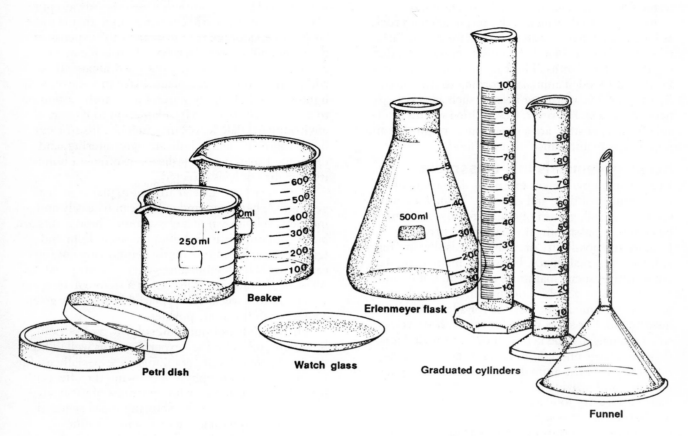

Beaker

Erlenmeyer flask

Graduated cylinders

Funnel

Petri dish

Watch glass

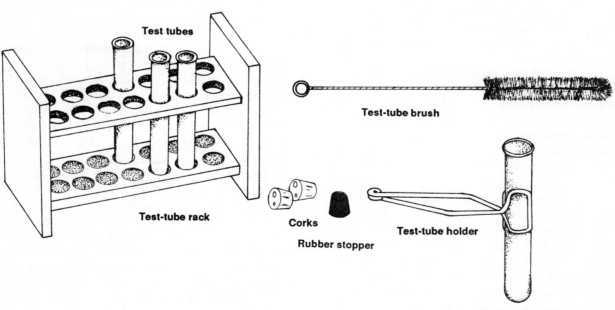

Test tubes

Test-tube rack

Test-tube brush

Corks

Rubber stopper

Test-tube holder

Laboratory Equipment

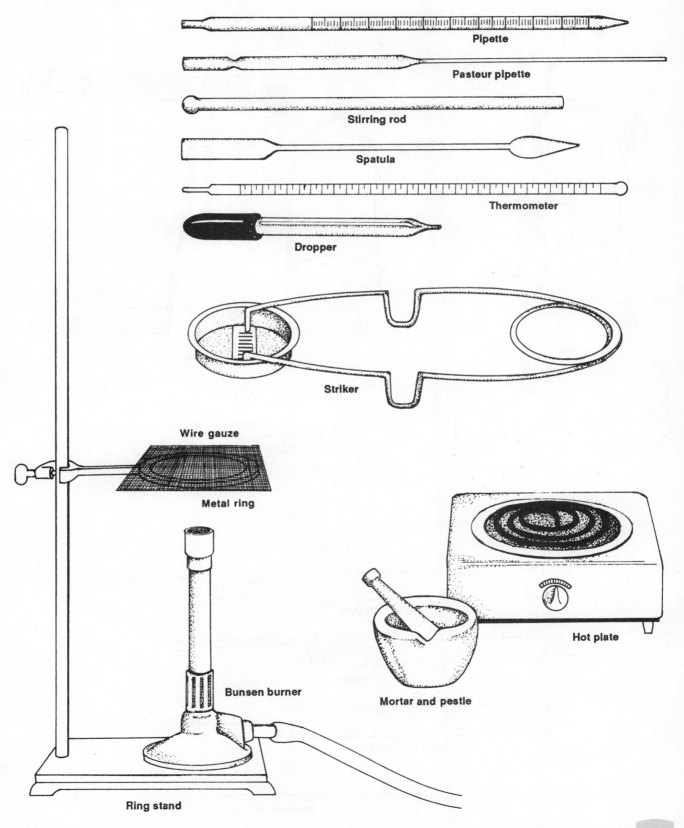

Pipette

Pasteur pipette

Stirring rod

Spatula

Thermometer

Dropper

Striker

Wire gauze

Metal ring

Hot plate

Bunsen burner

Mortar and pestle

Ring stand

Laboratory Equipment

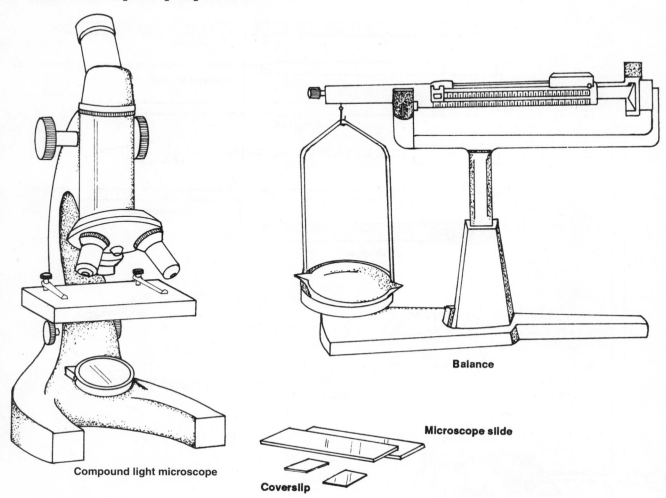

Compound light microscope

Balance

Microscope slide

Coverslip

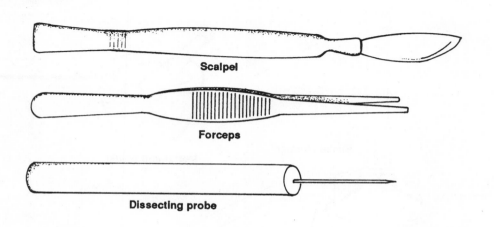

Scalpel

Forceps

Dissecting probe

Inoculating loop

Safety in the Laboratory

1. Always obtain your teacher's permission to begin a lab.

2. Study the procedure. If you have questions, ask your teacher. Be sure you understand all safety symbols shown.

3. Use the safety equipment provided for you. Goggles and a laboratory apron should be worn when any lab calls for using chemicals.

4. When you are heating a test tube, always slant it so the mouth points away from you and others.

5. Never eat or drink in the lab. Never inhale chemicals. Do not taste any substance or draw any material into your mouth.

6. If you spill any chemical, immediately wash it off with water. Report the spill immediately to your teacher.

7. Know the location and proper use of the fire extinguisher, safety shower, fire blanket, first aid kit, and fire alarm.

8. Keep all materials away from open flames. Tie back long hair and loose clothing.

9. If a fire should break out in the lab, or if your clothing should catch fire, smother it with the fire blanket or a coat, or get under a safety shower. **NEVER RUN.**

10. Report any accident or injury, no matter how small, to your teacher.

Follow these procedures as you clean up your work area.

1. Turn off the water and gas. Disconnect electrical devices.

2. Return materials to their places.

3. Dispose of chemicals and other materials as directed by your teacher. Place broken glass and solid substances in the proper containers. Never discard materials in the sink.

4. Clean your work area.

5. Wash your hands thoroughly after working in the laboratory.

First Aid in the Laboratory	
Injury	**Safe response**
Burns	Apply cold water. Call your teacher immediately.
Cuts and bruises	Stop any bleeding by applying direct pressure. Cover cuts with a clean dressing. Apply cold compresses to bruises. Call your teacher immediately.
Fainting	Leave the person lying down. Loosen any tight clothing and keep crowds away. Call your teacher immediately.
Foreign matter in eye	Flush with plenty of water. Use an eyewash bottle or fountain.
Poisoning	Note the suspected poisoning agent and call your teacher immediately.
Any spills on skin	Flush with large amounts of water or use safety shower. Call your teacher immediately.

Safety Contract

I, _Raneen Ayoub_ , have read and understand the safety rules and first aid information listed above. I recognize my responsibility and pledge to observe all safety rules in the science classroom at all times.

Raneen Ayoub
signature

_8/14/07_
date

Safety Symbols

The **Biology: The Dynamics of Life** program uses several safety symbols to alert you to possible laboratory dangers. These safety symbols are explained below. Be sure that you understand each symbol before you begin a lab activity.

SAFETY SYMBOLS	HAZARD	EXAMPLES	PRECAUTION	REMEDY
DISPOSAL	Special disposal procedures need to be followed.	certain chemicals, living organisms	Do not dispose of these materials in the sink or trash can.	Dispose of wastes as directed by your teacher.
BIOLOGICAL	Organisms or other biological materials that might be harmful to humans	bacteria, fungi, blood, unpreserved tissues, plant materials	Avoid skin contact with these materials. Wear mask or gloves.	Notify your teacher if you suspect contact with material. Wash hands thoroughly.
EXTREME TEMPERATURE	Objects that can burn skin by being too cold or too hot	boiling liquids, hot plates, dry ice, liquid nitrogen	Use proper protection when handling.	Go to your teacher for first aid.
SHARP OBJECT	Use of tools or glassware that can easily puncture or slice skin	razor blades, pins, scalpels, pointed tools, dissecting probes, broken glass	Practice common-sense behavior and follow guidelines for use of the tool.	Go to your teacher for first aid.
FUME	Possible danger to respiratory tract from fumes	ammonia, acetone, nail polish remover, heated sulfur, moth balls	Make sure there is good ventilation. Never smell fumes directly. Wear a mask.	Leave foul area and notify your teacher immediately.
ELECTRICAL	Possible danger from electrical shock or burn	improper grounding, liquid spills, short circuits, exposed wires	Double-check setup with teacher. Check condition of wires and apparatus.	Do not attempt to fix electrical problems. Notify your teacher immediately.
IRRITANT	Substances that can irritate the skin or mucous membranes of the respiratory tract	pollen, moth balls, steel wool, fiber glass, potassium permanganate	Wear dust mask and gloves. Practice extra care when handling these materials.	Go to your teacher for first aid.
CHEMICAL	Chemicals that can react with and destroy tissue and other materials	bleaches such as hydrogen peroxide; acids such as sulfuric acid, hydrochloric acid; bases such as ammonia, sodium hydroxide	Wear goggles, gloves, and an apron.	Immediately flush the affected area with water and notify your teacher.
TOXIC	Substance may be poisonous if touched, inhaled, or swallowed	mercury, many metal compounds, iodine, poinsettia plant parts	Follow your teacher's instructions.	Always wash hands thoroughly after use. Go to your teacher for first aid.
OPEN FLAME	Open flame may ignite flammable chemicals, loose clothing, or hair	alcohol, kerosene, potassium permanganate, hair, clothing	Tie back hair. Avoid wearing loose clothing. Avoid open flames when using flammable chemicals. Be aware of locations of fire safety equipment.	Notify your teacher immediately. Use fire safety equipment if applicable.

 Eye Safety Proper eye protection should be worn at all times by anyone performing or observing science activities.

 Clothing Protection This symbol appears when substances could stain or burn clothing.

 Animal Safety This symbol appears when safety of animals and students must be ensured.

 Radioactivity This symbol appears when radioactive materials are used.

INVESTIGATION

How Can Scientific Methods Be Used to Solve a Problem?

Scientific methods are steps scientists use to answer questions or solve problems. These steps usually include observation, hypothesis formation, experimentation, and interpretation of experimental results. Scientific methods often have been compared with the procedure a detective uses in solving a crime or problem.

OBJECTIVES

- Use scientific methods to decide whether two liquids are the same or different.
- Make careful observations.
- Hypothesize whether the two liquids are the same or different.
- Record accurate experimental results.

MATERIALS

Erlenmeyer flask with liquid, labeled A
Erlenmeyer flask with liquid, labeled B
clock or watch with second hand

stoppers to fit flasks (2)
beaker
laboratory apron
goggles

PROCEDURE

Part A. Observation

Accurate observations are a necessary part of scientific methods.

1. Examine the two flasks. DO NOT remove the stoppers and DO NOT shake the contents.

2. Record in Table 1 two or three similarities and differences between the contents of the two flasks.

   a. Do you think both flasks contain the same liquid? Explain.

   yes. Because they look the same

   b. Is your answer to question 2a based on experimentation or guessing?

   guessing

   c. Would a scientist guess an answer to a question or experiment first?

   ~~exped~~ question

   d. Do both flasks contain the same volume of liquid?

   _no_

   e. What gas might be in the upper half of flask A that is not in flask B?

   blue substance

   f. Is there any direct evidence for your answer to question 2e?

   yes

3. Make a hypothesis about whether the contents of the two flasks are the same or different. Write your hypothesis in the space provided.

Part B. Experimentation

In determining whether the two liquids are the same or not, a scientist would carry out some experiments. Experimentation is another part of scientific methods.

Experiment 1. What happens if you shake the liquids?

1. Give each flask *one hard shake using an up-and-down motion of your hand*. Make sure your thumb covers the stopper as you shake. See Figure 1. Observe each flask carefully.

2. Record your observations in Table 2. Again, look for similarities and differences in the contents of the two flasks.

How Can Scientific Methods Be Used to Solve a Problem?

PROCEDURE continued

a. After shaking the flasks, do you think they contain different liquids?

yes

b. What was present in flask A that might have been responsible for the change in the liquid?

shaking

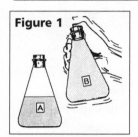
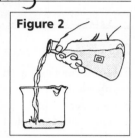

Figure 1

Figure 2

Experiment 2. What happens if you remove some of the liquid in flask B so it contains the same volume of liquid as flask A?

1. Remove the stopper from flask B and pour out half of the contents into a beaker. See Figure 2. Make sure that the volume of liquid remaining in flask B is equal to the volume of liquid in flask A. Replace the stopper.

2. Give both flasks *one hard shake using an up-and-down motion of your hand.* Hold the stopper in place while shaking.

3. Observe each flask carefully.

4. In Table 3, record any similarities or differences you observed.

a. Do both flasks now appear to contain the same liquid?

yes

b. What was added to flask B that was not present before?

 nothing

Experiment 3. What happens if you shake the flasks more than once?

1. Shake each flask *hard once with an up-and-down motion.*

2. Note the number of seconds it takes for each liquid to return to its original condition after shaking. Record the time under "1 shake, Trial 1" in Table 4.

3. Shake each flask *hard twice with an up-and-down motion.*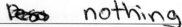

4. Time how long it takes for the liquids to return to their original conditions. Record your data under "2 shakes, Trial 1" in Table 4.

5. Shake both flasks *hard three times with an up-and-down motion.*

6. Note and record under "3 shakes, Trial 1" in Table 4 the time it takes for the liquids to return to their original conditions.

a. Is the time needed for the liquids in flasks A and B to return to their original conditions after one shake about the same? _____

b. Is the time needed for the liquids in flasks A and B to return to their original conditions after two or three shakes the same? _____

c. Look at your data in Table 4. Does flask A show an increase or decrease in time needed to return to its original condition as the number of shakes increases from one to three? _____

Does flask B show a similar change?

_____

7. In any experiment, running several trials reduces the probability of making errors. Run two more trials for each part of Experiment 3. Be sure to keep track of the amount of time needed for the liquids to return to their original conditions.

8. Record the results of Trials 2 and 3 in Table 4.

a. Do three trials give better evidence that the liquid in flask A is "behaving" in a way similar to the liquid in flask B after shaking each flask

once? _____

twice? _____

three times? _____

b. Do three trials give better evidence that an increase in time is needed for the liquid to return to its original condition as the number of shakes increases from one to three

for flask A? _____

for flask B? _____

How Can Scientific Methods Be Used to Solve a Problem?

Lab 1-1

HYPOTHESIS

DATA AND OBSERVATIONS

Table 1

First Observations	
Similarities	Differences

Table 2

Results of Experiment 1	
Similarities	Differences

Table 3

Results of Experiment 2	
Similarities	Differences

Table 4

Three Trials of Experiment 3								
Time in seconds to return to original condition								
1 shake			2 shakes			3 shakes		
Trial	1	2	3	1	2	3	1	2	3
Flask A		15	1004			1:04			
Flask B		18	1004			1:04			

How Can Scientific Methods Be Used to Solve a Problem?

ANALYSIS

Questions 1–4 should help you to interpret what you have observed. Interpretations are reasonings based on observations and experiments.

1. On the basis of your observations in Part A, could you decide if both flasks contained the same liquid?

_____

2. After performing Experiment 1, could you decide if both flasks contained the same liquid?

_____

3. Which experiment or experiments helped you to decide that the liquids in flasks A and B were the same or different? Explain.

_____

4. Besides the liquid itself, what else was needed for the liquid to change color?

_____

_____

Questions 5–7 should help you to form a conclusion.

5. Explain why flask B did not change color when shaken in Experiment 1.

_____

6. Why must the liquids in the half-filled flasks be shaken to produce a color change?

_____

7. Why did more shaking increase the amount of time needed for the liquids in flasks A and B to change back to their original color?

_____

8. a. Could you have answered the first question in Part A by guessing? _____

    b. Why is experimenting a better method of problem solving than guessing?

_____

    c. What is meant by the phrase "solving a problem by using scientific methods"?

_____

CHECKING YOUR HYPOTHESIS

Was your **hypothesis** supported by your data? Why or why not?

_____

_____

FURTHER INVESTIGATIONS

1. Design an experiment to test the effect of a new fertilizer on plant growth.

2. Design and carry out an experiment to test the effect of different colors of light on plant growth.

Lab 1-2

Using SI Units

Scientists all over the world use SI units for measuring. SI stands for the International System of Measurement. The use of SI units allows scientists from different countries to communicate easily.

The SI system is more convenient than the English system of inches, feet, ounces, pounds, and so on, because all SI units use a base with standard prefixes and units that are multiples of 10. For example, there are 100 centimeters in a meter and 1000 meters in a kilometer.

Converting from one SI unit to another involves multiplying or dividing by 10 or a multiple of 10. By memorizing a few standard prefixes and knowing when to multiply or divide, unit conversion is easy. When converting from a small unit to a larger unit, divide. When converting from a large unit to a smaller unit, multiply. The number you use to divide or multiply with depends on the units involved. For example, in converting from millimeters to centimeters, divide by 10 because there are 10 millimeters in one centimeter. SI units will be used in this Exploration to measure length, mass, and volume as you work with growing plants.

OBJECTIVES

- Use SI units to measure the mass of bean seeds.
- Use SI units to measure, record, and graph the growth of bean plants.

MATERIALS

bean seeds (12)   colored pencils (4)   newspaper   vermiculite
masking tape   beakers (4)   wax marking pencil   paper towels
balance   metric ruler   fertilizer solutions: laboratory apron
10-cm flower pots (4)   50-mL graduated   full-strength,   goggles
water   cylinder   double-strength,
                                                          half-strength

PROCEDURE

Part A. Seed Mass

1. Separate the bean seeds into four groups of three seeds each. Label the groups A, B, C, and D.

   A gram (g) is a common SI unit of mass. Other common units of mass are the milligram (mg) and the kilogram (kg). A milligram is 1/1000 of a gram. A kilogram is 1000 grams.

2. Using a balance, determine the average mass in grams of the seeds in each group. To do this, find the mass of the entire group and divide by the number of seeds. Record the total mass and the average mass in Table 1.

3. Use the information in the introduction to convert the average mass of the seeds in each group to milligrams and kilograms. Record these conversions in Table 1.

Part B. Plant Height

1. Using masking tape, label four flower pots A, B, C, and D. Add an 8-cm layer of vermiculite to the bottom of each pot. Plant Group A seeds in Pot A, Group B seeds in Pot B, and so on. Plant the seeds 2 cm deep.

2. Label four beakers as follows: A—tap water; B—plant fertilizer; C—double-strength plant fertilizer; and D—half-strength plant fertilizer. Pour some of each solution into the appropriate beaker.

   A liter (L) is a common SI unit of volume. Another common unit of volume is the milliliter (mL). A milliliter is 1/1000 of a liter.

3. Using a graduated cylinder, measure 50 mL of each solution and carefully water each pot with

Using SI Units

PROCEDURE continued

the corresponding solution. Wash the cylinder between solutions.

4. Continue to water all four pots with 50 mL of solution each time you water.

5. Observe the pots of seeds daily. On each pot's label, record the date when the first seedling emerges in that pot.

A centimeter (cm) is a common SI unit of length. Other units of length include the millimeter (mm), meter (m), and kilometer (km). A centimeter is 1/100 of a meter. A millimeter is 1/1000 of a meter. There are 1000 millimeters in a meter. A meter is 1/1000 of a kilometer. There are 1000 meters in a kilometer.

6. On the third day after the first seedling emerges in a pot, measure the height in centimeters of each plant in that pot. Determine plant height by measuring from the top of the vermiculite to the tip of the main plant stem, as in Figure 1. If a plant droops, gently straighten it to get its actual height. Record the plant heights in each pot in Table 2.

7. Determine the average height of the plants in each pot. Record these average heights in Table 2.

8. On the sixth day, measure, average, and record in Table 3 the heights of the plants in each pot.

9. On the ninth day, measure, average, and record in Table 4 the heights of the plants in each pot.

10. In Tables 2, 3, and 4, convert the average heights of the plants in each of the pots to millimeters, meters, and kilometers. Be sure to multiply or divide by the correct multiple of 10 to get the correct conversions.

11. Using a different colored pencil for each pot, graph the average height of the plants in pots A–D in Figure 2.

Part C. Plant Mass

1. After the ninth day, spread newspapers over your work area and carefully remove the plants and vermiculite from Pot A. Being careful not to break the roots, gently remove the vermiculite from around the roots. Remove any clinging material by rinsing the roots gently in tap water. Dry the roots with a paper towel.

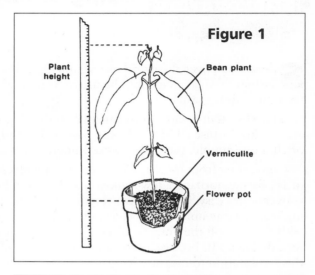

Figure 1

Plant height

Bean plant

Vermiculite

Flower pot

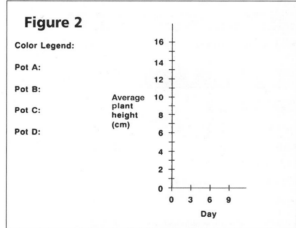

Figure 2

Color Legend:

Pot A:

Pot B:

Pot C:

Pot D:

Average plant height (cm)

2. Using a balance, determine and record in Table 5 the combined mass of the three plants in pot A. Then calculate the average mass of a plant in Pot A and record it in Table 5.

3. Repeat steps 1 and 2 for pots B, C, and D.

4. Record average seed masses from Table 1 in the proper column of Table 5.

5. Record the increase of mass in Table 5.

$$\text{increase of mass} = \text{average plant mass} - \text{average seed mass}$$

6. Determine the percentage increase in mass of the plants from each pot. Record the percentage increases in Table 5.

$$\text{percentage increase in mass} = \frac{\text{increase of mass}}{\text{average seed mass}} \times 100\%$$

Copyright © Glencoe/McGraw-Hill, a division of The McGraw-Hill Companies, Inc.

Using SI Units

DATA AND OBSERVATIONS

Table 1

Mass of Seeds				
Group	Total mass of seeds (g)	Average mass (g)	Milligrams (mg)	Kilograms (kg)
A				
B				
C				
D				

Table 2

Height of Plants 3 Days After Germination					
Pot	Height of each plant (cm)	Average height (cm)	Millimeters (mm)	Meters (m)	Kilometers (km)
A					
B					
C					
D					

Table 3

Height of Plants 6 Days After Germination					
Pot	Height of each plant (cm)	Average height (cm)	Millimeters (mm)	Meters (m)	Kilometers (km)
A					
B					
C					
D					

Table 4

Height of Plants 9 Days After Germination					
Pot	Height of each plant (cm)	Average height (cm)	Millimeters (mm)	Meters (m)	Kilometers (km)
A					
B					
C					
D					

Table 5

Plant Mass and Mass Increase				
Pot	Combined mass of plants (g)	Average plant mass (g)	Average seed mass (g)	Increase of mass (g)	Percentage increase
A					
B					
C					
D					

Using SI Units

ANALYSIS

1. What are the common SI units of mass? _____

Of length? _____

Of volume? _____

2. How do you convert:

   a. grams to kilograms? _____   **d.** centimeters to meters? _____

   b. grams to milligrams? _____   **e.** liters to milliliters? _____

   c. centimeters to millimeters? _____   **f.** milliliters to liters? _____

3. Why was it necessary to water each pot at the same time with the same amount of solution?

_____

_____

4. Rank the pots in order of growth from least average height to greatest average height. Explain your results.

_____

_____

_____

5. Rank the pots in order of percentage increase in plant mass from least increase to greatest increase.

_____

6. How does the ranking for mass compare with the ranking for plant growth?

_____

7. What are possible explanations for differences in the rankings?

_____

_____

_____

FURTHER EXPLORATIONS

1. Design and conduct an experiment to determine if bean seed size has any effect on the size and growth of bean plants.

2. Design and conduct an experiment to compare the relative effectiveness of several different types of commercial plant fertilizers.

Lab 2-1

EXPLORATION   Physical Factors of Soil

Soil is a major factor influencing the survival of many living things. Many organisms live in the soil. Others are anchored in soil and obtain water and minerals from it. Still other organisms depend on these soil-dependent organisms for food. The physical properties of a particular kind of soil determine the kinds of organisms that live in or on it.

OBJECTIVES

- Compare the amounts of various particle types in three soil samples.
- Calculate the water-holding capacities of three soil samples.

MATERIALS

specimen jars with lids (3)
soil samples, enough of each to fill a specimen jar (3)
20-cm cloth squares (3)
string
balance
beakers (3)
scoop

water
metric ruler
pins
masking tape
plastic or rubber gloves
laboratory apron
goggles

PROCEDURE

Part A. Particle Types

1. Label three specimen jars with the locations of the soil samples. Write the location in Table 1. Fill each jar halfway with soil. Add water, allowing it to soak into the soils, until the jars are full.

2. Cover the jars with lids and shake until the large soil particles break apart. Set the jars aside and let the particles settle overnight.

3. Using a ruler, measure in millimeters the depth of each particle type in each jar. See Figure 1.

4. Record in Table 1 the depths of the gravel, coarse sand, fine sand, silt, and clay layers in the settled soil samples.

Part B. Water-Holding Capacity

1. Place a scoop of soil in each cloth. Do not use soil from the jars in Part A. Wrap and tie the soil samples in the cloths. Use pins to attach labels identifying the samples.

2. Place the wrapped samples where they will dry completely, then determine and record their masses in Table 2.

3. Place each wrapped, dried soil sample in a beaker of water for five to ten minutes or until the soil is saturated.

Figure 1

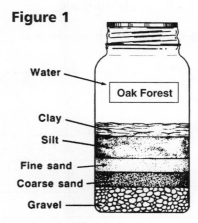

Water
Oak Forest
Clay
Silt
Fine sand
Coarse sand
Gravel

4. Remove the wrapped samples from the beakers and allow excess water to drain from them through the cloths. Find and record the masses of the saturated samples.

5. Calculate the water-holding capacity of each sample as a percentage of the dry mass.

$$\text{percentage water-holding capacity} = \frac{\substack{\text{mass of saturated} \\ \text{soil and cloth}} - \substack{\text{mass of dried} \\ \text{soil and cloth}}}{\substack{\text{mass of dried} \\ \text{soil and cloth}}} \times 100$$

Physical Factors of Soil

DATA AND OBSERVATIONS

Table 1

Soil Particle Type					
	Depth of each particle type (in mm)				
Soil location	Gravel	Coarse sand	Fine sand	Silt	Clay
1.					
2.					
3.					

Table 2

Water-Holding Capacity			
Soil location	Mass of dried soil and cloth	Mass of saturated soil and cloth	Percentage water-holding capacity
1.			
2.			
3.			

ANALYSIS

1. Which type of soil particle made up:

   a. the greatest amount of each soil sample? _____

   b. the least amount? _____

2. Which type of soil was:

   a. most closely packed? Explain. _____

   _____

   b. least closely packed? Explain. _____

   _____

3. In terms of water-holding capacity, what type of soil is best for growing plants? Explain.

   _____

   _____

FURTHER EXPLORATIONS

1. Research the procedure for calculating the organic-matter content of a soil sample.

2. Prepare a chart showing the predominant soil types in various parts of the United States. Show how soil types affect the commercial activities of an area.

INVESTIGATION

What Organisms Make Up a Microcommunity?

Usually, when a community and the organisms associated with it are described, observable organisms such as trees, grass, pigeons, and squirrels are listed. These observable organisms are called macroorganisms. However, a community also has many microorganisms. In fact, microorganisms can make up their own microcommunities. These microcommunities are small and inconspicuous. With a microscope, you can examine microcommunities that would ordinarily go undetected. It is also possible to identify and classify many of the organisms found in microcommunities.

OBJECTIVES

- Observe and identify the organisms found in the microcommunities in bean water and pond water.

- Determine if each organism is motile (able to move) or sessile (unable to move).

- Identify each organism as a producer or a consumer.
- Make a hypothesis about the relationship between consumers and producers and their motility.

MATERIALS

water
pond water
dropper
spring-type clothespin
bean water (tap water
    in which beans have
    soaked for several days)

microscope slides (3)
coverslips (2)
Bunsen burner
crystal violet stain
striker
small beaker or
    paper cup

100-mL graduated
    cylinder
filter paper
compound light
    microscope
funnel

250-mL beaker
plastic or rubber gloves
laboratory apron
goggles

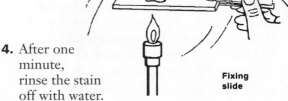

Figure 1

Fixing
slide

PROCEDURE

Part A. Bean-Water Microcommunity

1. Place a drop of bean water on a microscope slide, spreading it into a thin film the size of a nickel.

2. Attach a clothespin to one end of the slide. Holding the slide by the clothespin, quickly pass the slide with the bean-water film through a low Bunsen burner flame several times. See Figure 1. **CAUTION:** *Always be careful around open flames. Secure loose hair and clothing to keep them away from the flame.* Warm the slide until the bean water has evaporated, leaving a residue. This process is called "fixing." Fixing sticks the cells to the slide.

3. Add several drops of crystal violet stain to the dried film of bean water. See Figure 2. Staining the slide will make microorganisms on the slide easier to see.

4. After one minute, rinse the stain off with water. This rinsing is best done by dipping the slide into a beaker or paper cup filled with water. Handling the slide with the clothespin will keep you from staining your fingers. See Figure 3.

5. Allow your slide to air dry. After the slide has dried, examine the bean-water microcommunity under the high power of your microscope. No coverslip is needed.

6. Prepare a wet mount of bean water. This time a coverslip is needed. Examine the wet mount under the high power of your microscope.

What Organisms Make Up a Microcommunity?

PROCEDURE continued

7. Using the diagrams in Figure 5 for comparison, identify the organisms in your bean-water microcommunity. Note particularly the shapes of the organisms to aid in your identifications. Record in Table 1 the names of the organisms observed in the bean water.

8. By examining the wet mount, determine whether each organism is motile or sessile. Record your findings in Table 1.

9. By examining the wet mount, determine whether each organism is a producer or a consumer. Producers usually will contain colored pigments ranging from yellow to green to blue-green. Consumers usually are colorless. Record your findings in Table 1.

10. Make a **hypothesis** to explain the relationship between producers and consumers and their motility. Write your hypothesis in the space provided.

Part B. Pond-Water Microcommunity

1. Line a funnel with filter paper. Filter 100 mL of pond water into a beaker. See Figure 4. This procedure will concentrate any organisms present in the water and make them easier to find.

2. Remove the filter paper from the funnel after the last of the pond water has drained through. Turn the filter paper inside out and touch to a glass slide the moist end that used to be the tip of the paper cone.

3. Add a small drop of water to the slide, if it is dry, and add a coverslip. Observe this wet mount under low and high powers of your microscope. Using the diagrams in Figure 5 for comparison, identify the organisms in your pond-water microcommunity.

4. Record in Table 1 the names of the organisms observed in the pond water. Determine and record in Table 1 whether each organism is motile or sessile. Determine and record in Table 1 whether each organism is a producer or a consumer.

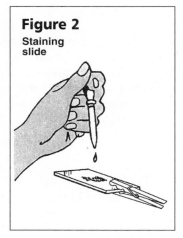

Figure 2
Staining slide

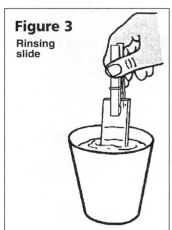

Figure 3
Rinsing slide

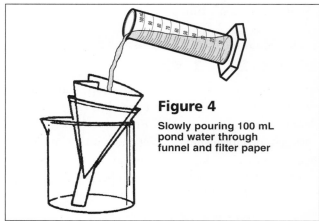

Figure 4
Slowly pouring 100 mL pond water through funnel and filter paper

HYPOTHESIS

_____

_____

_____

What Organisms Make Up a Microcommunity?

Figure 5

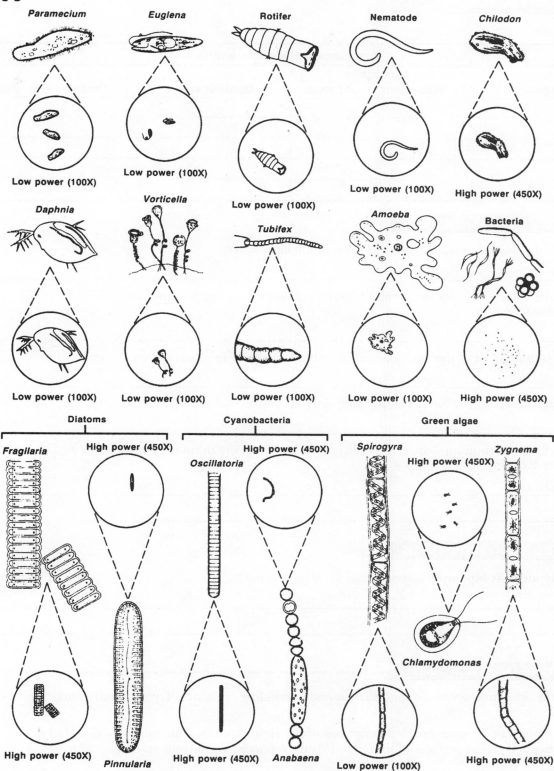

What Organisms Make Up a Microcommunity?

DATA AND OBSERVATIONS

Table 1

Organisms in Microcommunities				
	Community			
Name of organism	bean water	pond water	Motile or sessile	Producer or consumer

ANALYSIS

1. List several possible sources of the organisms in the bean water.

_____

2. Explain the color difference between producers and consumers in a pond-water microcommunity.

_____

3. a. Do you think most of the microorganisms in bean water are producers or consumers?

_____

   b. What evidence do you have that may support your answer?

_____

4. Did you find a relationship between motility and whether an organism is a consumer or a producer? Explain.

_____

_____

CHECKING YOUR HYPOTHESIS

Was your **hypothesis** supported by your data? Why or why not?

_____

_____

FURTHER INVESTIGATIONS

1. Examine other microcommunities, such as one in standing rainwater. Try to identify the organisms you see.

2. Read Anton van Leeuwenhoek's descriptions of the various microcommunities he observed in rainwater and on the surfaces of his teeth. Write a report based on your reading.

How Does the Environment Affect an Eagle Population?

With unlimited resources and ideal environmental conditions, a population would increase in size indefinitely. This rarely happens, however, because resources are limited and environmental conditions are not ideal. The carrying capacity of an ecosystem is the maximum number of organisms that the ecosystem can support. In nature, many populations remain below the carrying capacity because of a combination of both nonliving (abiotic) and living (biotic) factors. These factors include climate, habitat, available food, water supply, pollution, disease, and interactions between species, including predation, parasitism, and competition. The interactions between a population and the many components of an ecosystem are very complex. In this Investigation, you will make a simplified model of the effects of some abiotic and biotic environmental factors on a bald eagle population.

OBJECTIVES

- Make a model to show how abiotic and biotic factors affect a bald eagle population.

- Hypothesize how competition affects a bald eagle population.

MATERIALS

index card
20 cm × 20 cm pieces
 of paper (2)

metric ruler
scissors
graph paper

uncooked rice grains
 (300)
colored pencils (6 colors)

goggles

PROCEDURE

Part A. Graphing the Fish Catch

Eagles mate for life. Only one pair of eagles occupies, defends, and hunts a well-defined territory.

1. Mark off the edges of the two sheets of paper in centimeters. On each sheet, draw a grid of 400 1-cm squares as shown in Figure 1. Also draw a vertical and a horizontal axis that cross in the middle of the grid.

2. Cut two 1-cm squares from the index card. Label one square M for male and the other F for female.

3. Lay the two grids near each other on a flat surface. Scatter 150 grains of rice over one grid, which will represent a lake. Each rice grain represents a large fish in the lake. Eagles eat only the large fish.

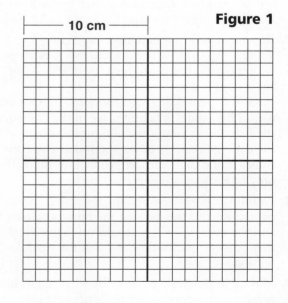

Figure 1

├── 10 cm ──┤

How Does the Environment Affect an Eagle Population?

PROCEDURE continued

Figure 2

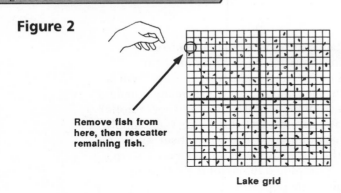

Remove fish from here, then rescatter remaining fish.

Lake grid

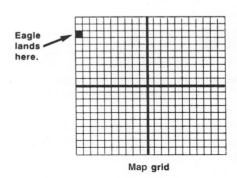

Eagle lands here.

Map grid

4. The other grid is a map of the lake. Hold the M (male eagle) square about 30 cm over the map grid and drop the M square onto the grid.

5. Note which grid square is most completely covered by the M square. Remove all the rice from the corresponding square on the lake grid, as shown in Figure 2. Do the same thing with the F square. This process represents the pair of eagles catching fish.

6. Each adult eagle hunts twice a day. Rescatter the remaining rice and repeat steps 4 and 5. Total the number of fish caught by each eagle on Day 1 and record the data for Day 1 in Table 1.

7. Repeat steps 4 to 6 nine times and complete Table 1. An eagle will share the fish it catches with its mate, but it will feed itself first. If an eagle does not eat a total of nine fish in any three-day period, it grows too weak to hunt and dies. This model is taking place in the fall when the fish do not reproduce. Be sure to examine the data for each three-day period as you continue. If one eagle dies, continue hunting with only one eagle for the remaining days.

8. Graph the data from Table 1, using a colored pencil. Record the days on the horizontal axis and the number of fish caught on the vertical axis. Answer questions 1 and 2 in the Analysis section.

9. Make a **hypothesis** to describe what might happen if two ospreys (other birds of prey) also hunted in the lake, each averaging a catch of

three fish per day. Write your hypothesis in the space provided.

10. Return all the rice to the lake grid. Repeat steps 4 to 7, but randomly remove an additional six fish per day that the ospreys catch. Record the data in Table 2.

11 Graph the data from Table 2, using a different colored pencil. Answer question 3 in the Analysis section.

Part B. Changing Abiotic and Biotic Factors

1. Rescatter all the rice on one grid, as before.

2. Other factors in the ecosystem affect the fish population and, thus, the eagle population. Follow directions for each factor described below. Record your data in Tables 3, 4, 5, and 6, labeling them properly. Be sure to rescatter the remaining rice before each hunting trip and scatter all the rice when you begin a new factor.

3. Graph the data from Tables 3 through 6. Use a different colored pencil for each table.

Factors That Might Affect an Ecosystem

A. A drought occurs that causes the water level of the lake to fall. This causes one-quarter of the fish to die. Remove 38 fish from the lake. Repeat steps 4 to 7 from Part A. Answer Analysis question 4.

B. It is spring. The fish in the lake are spawning. Double the number of fish (rice grains). Repeat steps 4 to 7 from Part A. Answer Analysis question 5.

How Does the Environment Affect an Eagle Population?

Lab 4-1

C. Phosphate pollution causes the algae in the lake to grow out of control. The algal growth reduces the amount of dissolved oxygen in the lake water and causes three-quarters of the fish to die. Remove 112 fish. Repeat steps 4 to 7 from Part A. Answer Analysis question 6.

D. The eagles have two offspring. The adults have to catch two additional fish each day. Repeat steps 4 to 7 of Part A. Answer Analysis question 7.

HYPOTHESIS

_____

_____

DATA AND OBSERVATIONS

Table 1

Day	1	2	3	4	5	6	7	8	9	10
Number of fish										

Table 2

Day	1	2	3	4	5	6	7	8	9	10
Number of fish										

Table 3

Day	1	2	3	4	5	6	7	8	9	10
Number of fish										

Table 4

Day	1	2	3	4	5	6	7	8	9	10
Number of fish										

Table 5

Day	1	2	3	4	5	6	7	8	9	10
Number of fish										

Table 6

Day	1	2	3	4	5	6	7	8	9	10
Number of fish										

Lab 4-1

How Does the Environment Affect an Eagle Population?

ANALYSIS

1. How does eagle predation affect the fish population over time?

_____

2. What effect, if any, might a small decrease in the fish population have on the eagle population?

_____

3. Ospreys and eagles compete for food. What effect, if any, might the competition have on the eagle population?

_____

_____

4. Explain how a climate change might or might not indirectly affect the eagle population.

_____

_____

5. What effect, if any, does an increase in the fish population have on the eagle population?

_____

6. Explain how phosphate pollution in a lake can indirectly affect the eagle population.

_____

7. How does an increase in the eagle population affect the fish population? How does this situation compare with the competition from ospreys?

_____

_____

CHECKING YOUR HYPOTHESIS

Was your **hypothesis** supported by your data? Why or why not?

_____

_____

FURTHER INVESTIGATIONS

1. Perform an experiment to test the effect of competition among bean plants. Germinate about 25 bean seeds. Select and plant one seedling in each of two pots three-quarters full of soil. These will serve as a control. Plant about 18 seedlings in another pot of the same size and filled with the same amount of soil. Measure the growth of the seedlings.

2. Design and carry out an experiment to test the effect of overcrowding in a yeast population. Since yeast reproduces by budding, a population increases rapidly. Observe changes in the number of yeast cells in the population by making a wet-mount slide and counting the number of cells in the field of view. How do light and temperature affect a yeast population?

EXPLORATION — The Lesson of the Kaibab

The environment may be altered by forces within the biotic community, as well as by interactions between organisms and the physical environment. The carrying capacity of an ecosystem is the maximum number of organisms that the ecosystem can support on a sustained basis. The changing density of a population may produce such profound changes in the environment that the environment becomes unsuitable for survival of that species. Humans can interfere with these natural interactions and have either a positive or a negative effect.

OBJECTIVES

- Graph the size of the Kaibab deer population of Arizona from 1905 to 1939.

- Analyze the actions responsible for the changes in the deer population.

- Propose a management plan for the Kaibab deer population.

MATERIALS

colored pencils (1 green and 1 red)

PROCEDURE

Before 1905, the deer on the Kaibab Plateau in Arizona were estimated to number about 4000 on almost 300 000 hectares of range. The average carrying capacity of the range was estimated then to be about 30 000 deer. On November 28, 1906, President Theodore Roosevelt created the Grand Canyon National Game Preserve to protect the "finest deer herd in America."

Unfortunately, by this time the Kaibab forest area had already been overgrazed by sheep, cattle, and horses. Most of the tall perennial grasses had been eliminated. The first step to protect the deer was to ban all hunting. In addition, in 1907, the Forest Service tried to exterminate the predators of the deer. Between 1907 and 1939, 816 mountain lions, 20 wolves, 7388 coyotes, and more than 500 bobcats, all predators of the deer, were killed.

1. Using the green pencil, draw and label a straight horizontal line across the graph in Data and Observations to represent the average carrying capacity of the range.

2. Using the red pencil, graph the data in Table 1.

3. Answer Analysis questions 1–4.

Signs that the deer population was out of control began to appear as early as 1920—the range was beginning to deteriorate rapidly. The Forest Service reduced the number of livestock-grazing permits. By 1923, the deer were reported to be on the verge of starvation, and the range conditions were described as "deplorable."

Table 1

Deer Population from 1905 to 1924	
Year	Deer population
1905	4 000
1910	9 000
1915	25 000
1920	65 000
1924	100 000

A Kaibab Deer Investigating Committee recommended that all livestock not owned by local residents be removed immediately from the range and that the number of deer be cut in half as quickly as possible. Hunting was reopened, and during the fall of 1924, 675 deer were killed by hunters. However, these deer represented only one-tenth the number that had been born that spring.

The Lesson of the Kaibab

Lab 4-2

PROCEDURE continued

4. Using the red pencil, plot the data in Table 2 on your graph. Label the completed graph.

5. Answer Analysis questions 5 and 6.

Today, the Arizona Game Commission carefully manages the Kaibab area with regulations geared to specific local needs. Hunting permits are issued to keep the deer in balance with their range. Predators are protected to help keep herds in balance with food supplies. Tragic winter losses can be checked only by keeping the number of deer near the carrying capacity of the range.

6. Answer Analysis questions 7–11.

Table 2

Deer Population from 1925 to 1939	
Year	**Deer population**
1925	60 000
1926	40 000
1927	37 000
1928	35 000
1929	30 000
1930	25 000
1931	20 000
1935	18 000
1939	10 000

DATA AND OBSERVATIONS

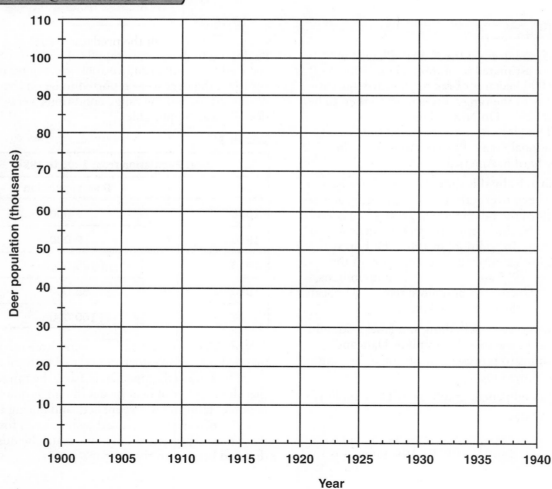

The Lesson of the Kaibab

ANALYSIS

1. In 1906 and 1907, what two methods did the Forest Service decide to use to protect the Kaibab deer?

_____

2. How many total predators were removed from the preserve between 1907 and 1939?

_____

3. What was the relationship of the deer herd to the carrying capacity of the range:

in 1915? _____

in 1920? _____

in 1924? _____

4. Did the Forest Service program appear to be successful between 1905 and 1924? Explain your answer.

_____

_____

_____

5. Why do you suppose the population of the deer declined in 1925 although the predators were being removed?

_____

_____

_____

6. Do you think any changes had occurred in the carrying capacity of the range from 1900 to 1940? Explain your answer.

_____

_____

_____

7. Why do you suppose the population of deer in 1905 was 4000 when the range had an estimated carrying capacity of 30 000?

_____

_____

_____

8. Without the well-meaning interference of humans, what do you think would have happened to the deer population after 1905?

_____

_____

_____

The Lesson of the Kaibab

Lab 4-2

ANALYSIS continued

9. What major lessons were learned from the Kaibab deer experience?

_____

_____

_____

10. If the lessons learned from the Kaibab deer studies had been known then, what recommendations would you have made in 1915?

_____

in 1923? _____

in 1939? _____

_____

11. What future management plan would you suggest for the Kaibab deer herd?

_____

_____

_____

FURTHER EXPLORATIONS

1. Many forests have been endangered by gypsy moth caterpillars. Research how they came to the United States and the methods that have been proposed to control them.

2. Mosquitoes are a great annoyance to many people. Obtain some books from your teacher or librarian that contain information on mosquitoes. Decide whether or not you would try to eliminate them. Justify your answer.

How Does Detergent Affect Seed Germination?

Synthetic detergents may contain phosphates and other chemicals that soften the water and prevent minerals and dirt from being redeposited onto the clothing during washing.

Phosphates are nutrients required by plants in small amounts. They exist naturally in clear lakes in very low concentrations. When many people began using phosphate detergents, problems resulted. The large amounts of phosphates dumped into bodies of water from household wastes greatly increased the concentration of phosphates in lakes and streams. The phosphates caused algal blooms, which decreased the oxygen level in the water. The balance of living things was disrupted, and ecological damage occurred.

Today, phosphates have been removed from many detergents to prevent such ecological damage. In fact, phosphate detergents are banned in areas around the Great Lakes and Chesapeake Bay. There are now, however, many additives in detergents that clean clothes the way phosphates do. These detergents without phosphates are referred to as biodegradable. Biodegradable substances can be decomposed by bacteria in water. Although these new detergents do not seem to cause major damage to the ecology, they may affect plants in other ways, such as influencing growth rate and seed germination.

OBJECTIVES

- Hypothesize what effects different concentrations of phosphate-free, biodegradable detergent have on seed germination and seedling growth.

- Determine the effects of different concentrations of phosphate-free, biodegradable detergent on seed germination and seedling growth.

- Graph the data collected in the experiment.

MATERIALS

masking tape
scissors
petri dishes (3)
wax marking pencil
toothpicks (3)

paper towels (3)
colored pencils (3)
radish seeds (90)
distilled water
50-mL graduated
    cylinder

graph paper
1% phosphate-free,
    biodegradable
    liquid dishwashing-
    detergent solution

10% phosphate-free,
    biodegradable
    liquid dishwashing-
    detergent solution
metric ruler
laboratory apron
goggles

PROCEDURE

1. Label the lids of three petri dishes CONTROL, 1%, and 10%. Label each dish with your name.

2. Cut six circles of paper toweling to fit the bottom of the dishes. Place two circles in the bottom of each dish as in Figure 1.

3. Distribute 30 radish seeds evenly over the bottom of each dish as in Figure 2.

4. Make sure the graduated cylinder is completely free of detergent by rinsing it thoroughly with

tap water, then with distilled water. Measure and pour 10 mL distilled water into the dish labeled CONTROL. Pour the water slowly and carefully so that the seeds are not disturbed.

5. Measure and carefully pour 10 mL of the 1% detergent solution into the dish labeled 1%.

6. Measure and carefully pour 10 mL of the 10% detergent solution into the dish labeled 10%. Rinse the graduated cylinder again with tap water, then with distilled water.

How Does Detergent Affect Seed Germination?

Figure 1

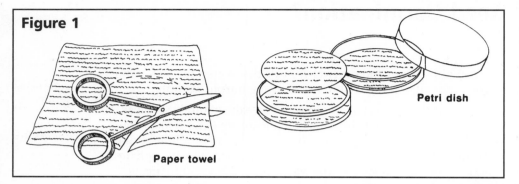

Paper towel

Petri dish

PROCEDURE continued

7. Use a toothpick to reposition your seeds, if necessary, so they are distributed evenly.

8. Replace the lids on the petri dishes and seal them with two short pieces of masking tape on opposite sides of the dish.

9. Place all the dishes in a warm, dark place.

10. Make a **hypothesis** about the effects of different concentrations of phosphate-free, biodegradable detergent on seed germination and seedling growth. Write your hypothesis in the space provided.

11. Examine your petri dishes daily for five days. Count the total number of seeds germinated each day in each dish. Germination has occurred if the root is visible. Record your counts in Table 1. Observe and measure the roots of some of the germinated seedlings as in Figure 3. Record your observations and measurements in Table 2.

12. Do not allow the paper towels in the petri dishes to dry out. Add small amounts of water or detergent solution of the proper concentration to the petri dishes if necessary. Be sure you add the same kind of liquid that you originally added to each dish.

13. Make a line graph of your data. Place the number of days on the horizontal axis and the number of germinated seeds on the vertical axis. For each treatment, plot the number of seeds germinated over the five-day period. Choose a different colored pencil for each of the three treatments.

Figure 2

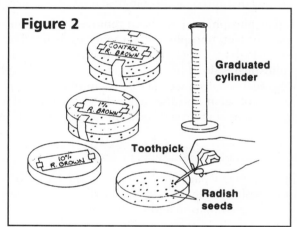

Graduated cylinder

Toothpick

Radish seeds

Figure 3

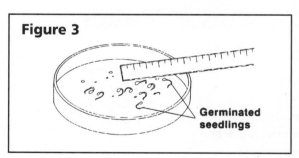

Germinated seedlings

HYPOTHESIS

_____

_____

_____

How Does Detergent Affect Seed Germination?

DATA AND OBSERVATIONS

Table 1

Number of Seeds Germinated			
Day	Control	1% detergent solution	10% detergent solution
1			
2			
3			
4			
5			

Table 2

Growth of Geminating Seedlings			
Day	Control	1% detergent solution	10% detergent solution

How Does Detergent Affect Seed Germination?

ANALYSIS

1. How many of the seeds germinated after five days

in distilled water? _____   in 1% detergent? _____   in 10% detergent? _____

2. How was germination of the seeds affected by the detergent?

_____

3. What was the purpose of the control?

_____

_____

4. What were the noticeable differences in growth of the seedlings in the three dishes
two days after germination?

_____

_____

5. Farmers often irrigate their crops with untreated water from lakes and streams. What
would happen to a farmer's crop yield if the water used for irrigation contained about
1% detergent?

_____

_____

_____

6. If a detergent is biodegradable, does that mean it will not harm living things? Explain.

_____

_____

_____

CHECKING YOUR HYPOTHESIS

Was your **hypothesis** supported by your data? Why or why not?

_____

_____

FURTHER INVESTIGATIONS

1. Experiment with a variety of detergent brands (including one with phosphates if available) and see if
they all have the same effects on seed germination and seedling growth.

2. Conduct the same experiment with more dilute and more concentrated detergents. Try to determine
what concentration of detergent, if any, gives the same results as the control. Try to determine what
concentration of detergent kills the seeds.

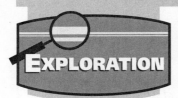

EXPLORATION   Tests for Organic Compounds

Understanding the chemistry of living organisms is an important part of biology. The structures of cells are made up of many different molecules. Cell metabolism involves the production and breakdown of many types of molecules. Most of the common molecules found in living things belong to four classes of carbon-containing molecules, or biomolecules: carbohydrates, lipids, proteins, and nucleic acids.

OBJECTIVES

- Determine the presence of starch by a chemical test.
- Analyze solutions for the presence of simple reducing sugars.

- Analyze a sample of vegetable oil for the presence of lipids.
- Analyze solutions for the presence of protein.

MATERIALS

droppers (9)
test tubes (3)
test-tube rack
test-tube holder
test-tube stoppers (2)

test-tube brush
brown paper
wax marking pencil
vegetable oil
biuret reagent
glucose solution

Benedict's solution
hot plate
water bath
soluble starch solution
water
95% ethanol

2% gelatin solution
iodine solution
clock or watch
thermal mitts (1 pair)
laboratory apron
goggles

PROCEDURE

Part A. Tests for Carbohydrates

Test for Starch

1. Put on goggles and an apron. Label three test tubes "1," "2," and "3," respectively. Place them in a test-tube rack.

2. Using a separate dropper for each solution, add 10 drops of soluble starch solution to test tube 1, 10 drops of glucose solution to test tube 2, and 10 drops of water to test tube 3. Record the color of each tube's contents in Table 1.

3. Add 3 drops of iodine solution to each test tube. **CAUTION:** *If iodine is spilled, rinse with water and call your teacher immediately.*

4. Record in Table 1 the color of each tube's contents after addition of the iodine. A blue-black color indicates the presence of starch.

5. Discard the contents of the test tubes according to your teacher's directions. Gently use a test-tube brush and soapy water to clean the three test tubes and rinse with clean water.

Figure 1

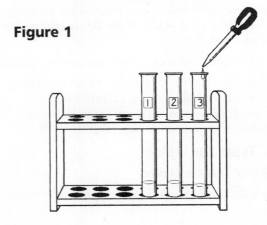

Tests for Organic Compounds

PROCEDURE continued

Test for Simple Reducing Sugars

1. Heat the water bath to boiling on the hot plate.

2. Label three test tubes "1," "2," and "3," respectively.

3. Using separate droppers for each solution, add 10 drops of soluble starch solution to test tube 1, 10 drops of glucose solution to test tube 2, and 10 drops of water to test tube 3. Record the color of each tube's contents in Table 2.

4. Add 20 drops of Benedict's solution to each of these three test tubes and place in a boiling water bath for 3 minutes. **CAUTION:** *If Benedict's solution is spilled, rinse with water and call your teacher immediately.*

Benedict's solution tests for the presence of simple reducing sugars (monosaccharides and some disaccharides, but not polysaccharides). Thus, a color change might or might not occur when Benedict's solution is added to a carbohydrate and heated. A change from blue to green, yellow, orange, or red occurs if a monosaccharide or certain disaccharides are present. The original blue color will remain after heating if a polysaccharide or certain other disaccharides are present.

Figure 2

5. Remove the three test tubes from the water bath using a test-tube holder and place them in a test-tube rack to cool. **CAUTION:** *Be careful not to burn yourself.*

6. Record the color of each tube's contents in Table 2.

7. Discard the contents of the test tubes according to your teacher's directions. Gently use a test-tube brush and soapy water to clean the three test tubes and rinse with clean water.

Part B. Tests for Lipids

Brown Paper Test for Lipids

1. Place a drop of water on a small piece of brown paper. Place a drop of oil on the same piece of paper. Allow the paper to dry for a few minutes.

2. Hold the piece of paper up to the light. If a semi-transparent (translucent) spot is evident, the sample contains lipids. Record the appearance of each spot in Table 3.

Solubility Test for Lipids

1. Label two test tubes "1" and "2," respectively.

2. Using separate droppers, add 20 drops of 95% ethanol to test tube 1 and 20 drops of water to test tube 2.

3. Add 5 drops of oil to test tubes 1 and 2 and stopper each tube.

4. Shake each tube well, let settle, and record in Table 4 whether the oil is soluble in either solvent.

Lipids are soluble only in nonpolar solvents because lipids, themselves, are nonpolar. Water is polar; ethanol is not.

5. Dispose of the contents of the test tubes according to your teacher's directions. Gently use a test-tube brush and soapy water to clean the two test tubes and rinse with clean water.

Part C. Tests for Proteins

1. Label three test tubes "1," "2," and "3," respectively.

2. Using separate droppers, add 30 drops of 2% gelatin to test tube 1, 30 drops of glucose solution to test tube 2, and 30 drops of water to test tube 3. Record the color of each tube's contents in Table 5.

3. Add 10 drops of biuret reagent to each test tube. **CAUTION:** *Biuret reagent is extremely caustic to the skin and clothing. If biuret reagent is spilled, rinse with water and call your teacher immediately.*

When biuret reagent is mixed with a protein, it will produce a lavender to violet color.

4. Record in Table 5 the color of each tube's contents after adding biuret reagent.

5. Discard the contents of the test tubes according to your teacher's directions. Gently use a test-tube brush and soapy water to clean the test tubes and rinse with clean water. Wash your hands.

6. Fill in the last column of all five tables with the correct interpretations of the test results.

Tests for Organic Compounds

DATA AND OBSERVATIONS

Table 1

Test for Starch				
Test tube	Substance	Color at start	Color after adding iodine	Starch present (+/−)
1	Starch			
2	Glucose			
3	Water			

Table 2

Test for Simple Reducing Sugars				
Test tube	Substance	Color at start	Color after adding Benedict's solution	Reducing sugar present (+/−)
1	Starch			
2	Glucose			
3	Water			

Table 3

Brown Paper Test for Lipids		
Substance	Translucent on brown paper?	Lipids present (+/−)
Water		
Oil		

Table 4

Solubility Test for Lipids			
Test tube	Substance	Dissolves? (yes/no)	Lipids present (+/−)
1	Oil in ethanol		
2	Oil in water		

Table 5

Test for Proteins				
Test tube	Substance	Color at start	Color after adding biuret reagent	Protein present (+/−)
1	Gelatin			
2	Glucose			
3	Water			

Tests for Organic Compounds

ANALYSIS

1. What is used to test for the presence of starch?

_____

_____

2. How can you tell by using this test that a substance contains starch?

_____

_____

3. What is used to test for the presence of simple reducing sugars such as monosaccharides?

_____

4. How can you tell by using this test that a substance contains a simple reducing sugar?

_____

_____

_____

5. Why was water tested for each chemical?

_____

_____

_____

6. What is used to test for the presence of protein?

_____

7. How can you tell by using this test that a substance contains protein?

_____

_____

8. Biuret reagent will turn the skin brownish-purple. Explain why this occurs.

_____

_____

_____

9. a. When greasy food is spilled on clothing, why is it difficult to clean with water alone?

_____

   b. What would be better than plain water for removing a greasy food stain? Why?

_____

FURTHER EXPLORATIONS

1. Choose a number of common substances that are available and conduct your own tests for the presence or absence of carbohydrates, lipids, and proteins. Cotton, animal hair, fingernails, and various foods such as egg white and cheese are among the many things you might test.

2. Investigate the difference between saturated and unsaturated fatty acids. Find recent information concerning the relationship between these fatty acids and good health.

Lab 6-2

What Is the Action of Diastase?

Enzymes are biological molecules that change the rates of reactions without being permanently changed or used up by the reactions themselves. Every living tissue and its cells depend upon enzymes for all cell reactions. Seeds contain enzymes that digest stored starch and make energy available for the embryo during germination. One seed enzyme, diastase, breaks down starch into the disaccharide maltose early in the germination process. The iodine and Benedict's tests can be used to confirm the activity of the enzyme diastase.

OBJECTIVES

- Prepare an extract of germinating barley seeds.
- Test known substances for the presence of starch and sugars.

- Hypothesize what changes will occur when barley extract is mixed with starch solution.
- Determine the action of barley extract on starch solution over a period of time.

MATERIALS

droppers (4)
test tubes (7)
test-tube rack
germinating barley
    seeds, 5 days old
    (25 seeds)
hot plate

water bath
Benedict's solution
0.4% starch solution
iodine solution
distilled water
0.1–0.2% diastase
    solution

mortar and pestle
100-mL beaker
50-mL beaker
clean sand
stirring rods (2)
scoop
plastic spot plate

test-tube holder
10-mL graduated
    cylinders (3)
white paper
clock or watch
thermal mitts (1 pair)
laboratory apron
goggles

PROCEDURE

Part A. Enzyme Preparation

1. Place a small scoop of clean sand in a mortar. Add all the germinating barley seeds and 10 mL distilled water and, using the pestle, grind the seeds until they are fine particles, as shown in Figure 1.

2. Pour the contents of the mortar into a 100-mL beaker. Add 40 mL distilled water and allow the mixture to settle for 15 to 20 minutes.

3. Save the barley extract you have made for use in Part B.

4. Set up and begin heating the water bath for Part B.

Part B. Testing Enzyme Activity

Iodine is used to test for the presence of starch. If starch is present, a blue-black color appears. The Benedict's test is used to test for some sugars such as

Figure 1

glucose and maltose. A positive reaction is shown by a color change from blue to orange.

1. Set a clean spot plate on a piece of white paper.

2. Add 3 drops of starch solution to a spot on the spot plate and 3 drops of diastase solution to a second spot. Add 1 drop of the iodine solution to each spot. See Figure 2. **CAUTION:** *If iodine solution is spilled, rinse with water and call your teacher immediately.*

What Is the Action of Diastase?

PROCEDURE continued

3. Note the color of the mixtures and record the presence or absence of starch in the proper places in Table 1.

4. Place 20 drops of starch solution into a test tube labeled "STARCH" and 20 drops of diastase solution into a test tube labeled "DIASTASE." Add 2 mL of Benedict's solution to each tube. Mix with two separate stirring rods or by gently tapping each tube against your hand. **CAUTION:** *If Benedict's solution is spilled, rinse with water and call your teacher immediately.*

5. Use the test-tube holder to place the test tubes into the boiling water bath for 3 minutes. **CAUTION:** *Be careful not to burn yourself.*

6. Remove the test tubes from the water bath using the test-tube holder and place them in the test-tube rack. Note the color in each tube. Record the presence or absence of sugars in the proper places in Table 1.

7. Make a **hypothesis** about the changes that will occur when barley extract is mixed with starch solution. Write your hypothesis in the space provided.

8. Label 5 test tubes "0 min," "3 min," "6 min," "9 min," and "12 min," respectively.

9. Using the graduated cylinder, place 10 mL of the barley extract in a 50-mL beaker and add 10 mL of starch solution. Be careful not to pour off any of the sand.

10. Immediately place 3 drops of this mixture from the beaker on a clean spot on the spot plate and test it for the presence of starch using the iodine solution. Record the results next to "0 min" in Table 2.

11. Quickly place 20 drops of the barley extract–starch mixture from the 50-mL beaker into the clean test tube labeled "0 min." Add 2 mL of Benedict's solution to the test tube and, using the test-tube holder, place the test tube in the water bath for 3 minutes. Record the results of this test next to "0 min" in Table 2.

Figure 2

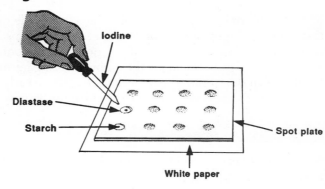

Figure 3

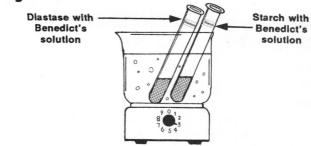

12. Repeat steps 10 and 11 every 3 minutes for 12 minutes (times 3, 6, 9, and 12 minutes). Place the mixture in the appropriately labeled test tube. For each repetition, record the results of the two tests in the appropriate places in Table 2.

13. Be sure to wash your hands when cleanup is complete.

HYPOTHESIS

_____

_____

_____

What Is the Action of Diastase?

Figure 4

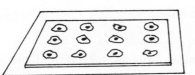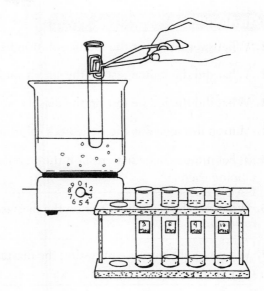

DATA AND OBSERVATIONS

Table 1

Testing for Substances		
Substance	Iodine test Starch (+/−)	Benedict's test Sugars (+/−)
Starch solution		
Diastase solution		

Table 2

Testing for Enzyme Action		
Time (minutes)	Iodine test Starch (+/−)	Benedict's test Sugars (+/−)
0		
3		
6		
9		
12		

What Is the Action of Diastase?

ANALYSIS

1. What substance was the iodine solution used to test for? _____

2. What did the iodine test on the starch solution indicate? _____

3. What did the iodine test on the diastase solution indicate? _____

4. What substance was the Benedict's solution used to test for? _____

5. What did the Benedict's test on the starch solution indicate? _____

_____

6. What did the Benedict's test on the diastase solution indicate?

_____

7. What was the purpose of testing the diastase solution with the iodine and the Benedict's solution?

_____

8. What did the two chemical tests indicate was in the barley extract–starch mixture at Time 0?

_____

9. a. Over the 12-minute period, what did the iodine test indicate was happening in the barley extract–starch mixture?

_____

_____

b. What did the Benedict's test indicate?

_____

_____

10. What happened to the starch in the barley extract–starch mixture?

_____

_____

CHECKING YOUR HYPOTHESIS

Was your **hypothesis** supported by your data? Why or why not?

_____

_____

FURTHER INVESTIGATIONS

1. Test for the activity of another common enzyme. Substitute the enzyme amylase, a component of human saliva, for diastase in this experiment. Amylase converts starch to maltose.

2. Design experiments to test the effects of boiling, freezing, dilution, or pH on the activity of diastase.

EXPLORATION

Use of the Compound Light Microscope

Possibly the most important instrument used by biologists is the microscope. A microscope aids scientists by allowing them to investigate worlds that otherwise are too small to be seen. A compound light microscope magnifies objects up to approximately 1500 times their natural size.

Two types of slides are used with the compound light microscope: prepared slides and temporary wet mounts. Prepared slides are made to last permanently. As their name suggests, temporary wet mount slides are not permanent. Many of the slides you will use in this course will be wet mounts. You will make these slides yourself.

OBJECTIVES

- Practice proper handling and use of the compound light microscope.
- Identify the parts of a compound light microscope.
- Locate objects under low- and high-power magnifications.
- Prepare a wet mount of an insect leg.

MATERIALS

compound light
  microscope
coverslip
preserved insect leg
lamp (if needed)
microscope slide
dropper

lens paper
water
forceps
plastic or rubber gloves
laboratory apron
goggles

PROCEDURE

Part A. Learning Microscope Parts and Functions

1. Look at Figure 1 to learn how to carry a microscope. Note that the student is carrying the microscope with two hands and holding it against her body. Also note that the microscope is carried straight up.

2. Position the concave (curved) surface of the mirror so that it is turned toward a light source, such as ceiling lights, windows, or a desk lamp. The mirror is attached to most microscopes by means of a swivel joint. If a lamp is built into your microscope, it replaces the mirror and outside light source. **CAUTION:** *Never use direct sunlight as a light source. Direct sunlight will damage your eyes.*

3. Look at Figure 2. Use the diagram that looks more like your microscope to locate microscope parts.

   a. Does your microscope have a lamp or a mirror?

   _____

   b. What type of diaphragm does your microscope have?

   _____

A diaphragm controls the amount of light entering the microscope. Turning the diaphragm adjusts the amount of light passing through the microscope.

Figure 1

4. Use Figure 2 to help you locate the revolving nosepiece, high-power objective, and low-power objective on your microscope. The low-power objective is identified by a 10× marking or by its short length. The high-power objective usually has a 43× marking and often is longer than the low-power objective. The objectives can be changed by turning the nosepiece as shown in Figure 3.

5. Place a check mark in the square next to each part of the microscope you have located.

   ❏ diaphragm   ❏ high-power objective
   ❏ lamp or mirror   ❏ low-power objective
   ❏ revolving nosepiece

Use of the Compound Light Microscope

Lab 7-1

Figure 2

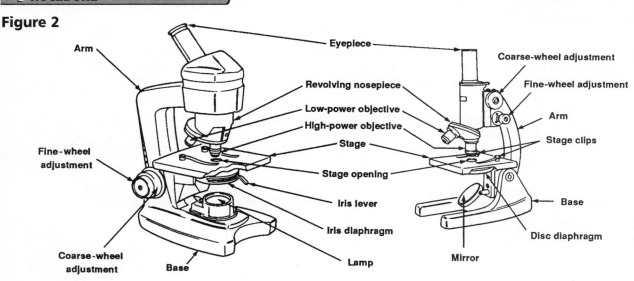

Do not continue with the lab until you know where these five parts are located.

6. Use Figure 2 to help you locate the eyepiece, coarse-wheel adjustment, fine-wheel adjustment, stage, and stage opening on your microscope.

7. Place a check mark in the square next to each part of the microscope you have located.

❑ eyepiece   ❑ stage

❑ coarse-wheel adjustment   ❑ stage opening

❑ fine-wheel adjustment

Do not continue with the lab until you know where these five parts are located.

Part B. Using the Microscope

1. Turn on the lamp or position the mirror toward the light source.

2. Turn and click the low-power objective so that it is directly over the stage opening. An objective is

Figure 3

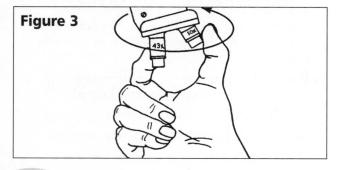

in proper viewing position when it is directly over the stage opening. Most microscopes will "click" when the objective is in proper viewing position.

3. Look through the eyepiece of the microscope. A circle of bright light should now be visible. Keep both eyes open. Keeping both eyes open will reduce eyestrain.

4. Adjust the mirror and diaphragm to make the circle of light as bright as possible.

5. Look to the side of the microscope as shown in Figure 4. Slowly turn the coarse-wheel adjustment back and forth. DO NOT force the wheel once it stops. When the wheel stops, turn it in the opposite direction. Note the movement of the low-power objective in relation to the stage.

a. In which direction does the objective move as you turn the coarse-wheel adjustment toward you? _____

b. In which direction does the objective move as you turn the coarse-wheel adjustment away from you? _____

6. The objectives and eyepiece should be cleaned with lens paper at the beginning of each laboratory period. Use one piece of paper and gently wipe each lens. *Always use lens paper to clean lenses. Other types of paper may scratch or smear lenses.*

Use of the Compound Light Microscope

> **PROCEDURE** continued

Part C. Preparation of a Temporary Wet Mount

A temporary wet mount consists of some object, such as an insect leg, placed in a drop of water on a slide with a coverslip over the object. Use the following steps in preparing your wet mount.

1. Add a small drop of water to a slide as shown in Figure 5A.

2. Place the insect leg to be viewed in the water drop. Use one insect leg only.

3. Use forceps to position a coverslip as shown in Figure 5B. Use of forceps prevents fingerprints getting on the coverslip.

4. Lower the edge of the coverslip down slowly over the water drop and object. This procedure will prevent the trapping of air under the coverslip.

Part D. Locating an Object Under the Microscope

1. Click the low-power objective into viewing position. **NOTE:** *Always locate an object first with low-power magnification even if a higher magnification is required for better viewing.*

2. Place the wet mount of the insect leg on the stage of your microscope. Position the slide on the stage so the insect leg is directly over the center of the stage opening. Secure the slide in place with the clips.

3. Look to the side of your microscope as shown in Figure 4. Slowly lower the low-power objective by turning the coarse-wheel adjustment until the objective almost touches the glass slide. Some microscopes have an automatic stop that prevents lowering the objective onto the slide. Other microscopes do not. **CAUTION:** *Never lower the objective toward the stage while looking through the eyepiece.*

4. While looking through the eyepiece with both eyes open, slowly turn the coarse-wheel adjustment so the objective rises, or moves away, from the stage. The insect leg should soon come into view.

5. Bring the insect leg into sharp focus by turning the fine-wheel adjustment.

Figure 4

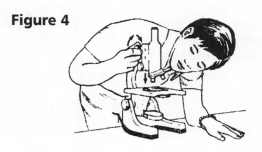

Figure 5

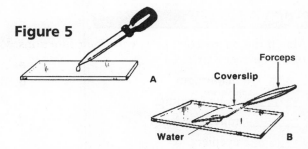

Part E. Increasing the Magnification of the Microscope

1. Any object to be viewed under high-power magnification is *always located first under low power and focused.* Locate and center the insect leg under low power of your microscope.

2. Move the low-power objective out of viewing position. Look first to the side of the microscope and then turn the nosepiece. Click the high-power objective into viewing position.

3. Look through the eyepiece. The insect leg should be visible. However, it may need to be focused. Use only the fine-wheel adjustment to sharpen the focus. **CAUTION:** *Never use the coarse-wheel adjustment for focusing with high power. Damage to the lens and slide may result if the coarse-wheel adjustment is used.*

4. If you are unable to find the insect leg, do the following: while looking through the eyepiece, move the glass slide slightly to the left, right, away from, or toward you. These movements may help to reposition the insect leg directly in the center of the high-power objective.

5. Repeat Parts D and E if you are unable to locate the object under high power.

Use of the Compound Light Microscope

DATA AND OBSERVATIONS

The eyepiece contains a glass lens that magnifies 10 times (10×). The low-power objective also contains a lens that magnifies 10 times (10×). Therefore, the total magnification of an object viewed under low power is 100×. Total magnification is calculated by multiplying the magnification of the objective by that of the eyepiece.

1. What is the total magnification of your microscope under low power? (Use the numbers printed on your low-power objective and eyepiece if present.) _____

2. What is the total magnification of your microscope under high power? (Use the numbers printed on your high-power objective and eyepiece.) _____

ANALYSIS

1. Match the microscope parts with their functions. Write the letter of the function in front of the correct part.

_____ diaphragm	**a.** Allows light to pass through stage
_____ stage opening	**b.** Brings objects into rapid but coarse focus
_____ mirror or lamp	**c.** Regulates amount of light entering microscope
_____ eyepiece	**d.** Is attached to revolving nosepiece and contains a lens capable of 10× magnification
_____ low-power objective	**e.** Contains a lens capable of 43× magnification
_____ high-power objective	**f.** Supports slide
_____ revolving nosepiece	**g.** Directs light into microscope
_____ coarse-wheel adjustment	**h.** Turns to change the objective positioned over the stage opening
_____ fine-wheel adjustment	**i.** Contains a lens capable of 10× magnification
_____ stage	**j.** Brings objects slowly into fine focus

2. Identify the following statements as true or false.

_____ **a.** Total magnification of a microscope is determined by adding the eyepiece-lens magnification to the objective-lens magnification.

_____ **b.** The fine-wheel adjustment must be used to sharpen focus when using high-power magnification.

_____ **c.** Always look to the side of a light microscope when lowering the objective.

_____ **d.** The eyepiece of a microscope is marked 10×. The high-power objective is marked 50×. The total magnification is 500×.

FURTHER EXPLORATIONS

1. Collect standing water from a variety of sources such as tree stumps, bird baths, or puddles. With a microscope, see how many different organisms you can find in the water. Check your library for books that might help in identifying the microscopic organisms.

2. Research the construction of early microscopes, such as the simple instruments made by Anton van Leeuwenhoek. Determine how a simple microscope is different from a compound light microscope and draw diagrams showing the optics of each.

INVESTIGATION — How Can a Microscope Be Used in the Laboratory?

Several important techniques and ideas can be mastered in the use of a compound light microscope. Knowing how and why your microscope works will enable you to make better observations. The techniques and hints presented in this Investigation will help you use your microscope correctly.

OBJECTIVES

- Compare the position of an object when viewed through a compound light microscope with its position on the microscope stage.
- Use stains to aid in viewing objects.

- Hypothesize how the field of view under low power compares with that under high power.
- Compare the depth of field under low and high powers.

MATERIALS

compound light microscope	absorbent cotton
microscope slide	single-edged razor blade
coverslip	black thread
scissors	white thread
dropper	tissue paper
water	peeled potato
lens paper	pond water
magazine page	plastic or rubber gloves
forceps	laboratory apron
iodine solution	goggles

PROCEDURE

Part A. Position of Objects When Viewed with a Microscope

1. Prepare a wet mount (see Exploration 7-1) of a lowercase letter "e" from a magazine page.

2. Place the wet mount of the letter "e" onto your microscope stage. Position the slide on the stage so the "e" faces you as it would on a magazine page.

3. Observe the letter "e" using low power on your microscope. (Review the procedure in Exploration 7-1 if necessary.) Focus the "e" with the fine adjustment. How does the orientation of the "e" viewed through the eyepiece compare with its orientation on the stage?

_____

4. While looking through the eyepiece, move the slide slowly from left to right. In what direction does the "e" move as seen through the microscope?

_____

5. While looking through the eyepiece, move the slide slowly toward you. In what direction does the "e" move when viewed through the microscope?

_____

Part B. Use of the Diaphragm

1. Prepare a wet mount of a few strands of absorbent cotton. (Follow the procedure outlined in Exploration 7-1.)

2. Observe the cotton fibers under low power. While looking through the microscope, change the amount of light entering the microscope by adjusting the diaphragm. Under what diaphragm setting (maximum, medium, or little light) are the cotton fibers sharpest?

_____

3. Change to high power and observe the cotton fibers. Again, readjust the amount of light entering the microscope. Under what diaphragm setting do the cotton fibers appear sharpest?

How Can a Microscope Be Used in the Laboratory?

PROCEDURE continued

Part C. Depth of Field

1. Cut a very short strand of black thread and of white thread.

2. Add a drop of water to a microscope slide. Cross the two strands to form an X in the water drop before adding a coverslip.

3. Locate the strands under low power. Center the slide so you are looking at the point where the strands cross. Adjust the diaphragm for proper lighting. Can both strands be observed clearly at the same time under low power?

_____

4. Change to high power and observe both strands at the point where they cross. Can both strands be observed clearly at the same time under high power?

_____

5. The lens system of your microscope allows you to see clearly only one depth at a time under high power. In order to see objects at different depths, turn the fine-wheel adjustment back and forth by a quarter of a turn while looking through the microscope. This movement will give a three-dimensional view of the object. Try this technique while looking at the crossed threads. Using Figure 1 as a guide, note that first one strand is in focus, then the other.

Part D. An Aid to Finding Proper Depth

Locating the proper depth of a wet mount is not always easy. Proper depth is especially hard to find when attempting to view moving organisms or when locating objects so small that they cannot be seen on the slide with the unaided eye. It is easy to focus at the wrong depth of the wet mount and waste valuable time looking at the top surface of the coverslip. The following technique will help you locate the proper depth of a wet mount.

1. Use a dropper to transfer a drop of pond water to a glass slide.

2. Add a hair from your head to the pond water. Add a coverslip.

3. Observe the wet mount under low power by first locating the strand of hair. Focus the hair and

Figure 1

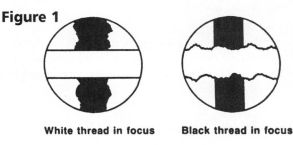

White thread in focus   Black thread in focus

Figure 2

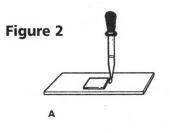

adjust the light. The slide can now be moved to find organisms in the pond water. No further adjustment is necessary with the coarse adjustment. You are at the proper depth for finding organisms because the hair and organisms are in the same plane.

4. Locate organisms in the pond water. Attempt to follow a moving organism.

Part E. Stains as an Aid to Microscope Work

Many objects observed with a microscope are colorless. Thus, they appear almost transparent and are difficult to see. Stains often are used in microscope work to color objects for easier and more detailed observation. Stains can be added to a wet mount without disturbing the slide.

1. With a single-edge razor blade, gently scrape the edge of a peeled potato. **CAUTION:** *Scrape away from your fingers.*

2. Add a drop of water to a glass slide. Mix the potato scrapings with the water. Add a coverslip.

3. View the wet mount under low power. You are looking at starch grains.

4. Diagram several grains in the circle marked "Unstained" in Data and Observations.

How Can a Microscope Be Used in the Laboratory?

5. Remove the slide from the microscope.

6. Add a drop of iodine solution to your slide along one edge of the coverslip as shown in Figure 2. Do not get any iodine on top of the coverslip. **CAUTION: *If iodine spillage occurs, wash with water and call the teacher immediately.***

7. Place a piece of tissue paper along the edge of the coverslip opposite the iodine solution. Allow the tissue paper to touch the water of the wet mount as shown in Figure 2. Water will soak into the tissue paper, drawing the iodine stain under the coverslip and into contact with the starch grains.

8. Observe the stained starch grains under low power.

9. Diagram several stained grains in the circle marked "Stained" in Data and Observations.

Part F. A Comparison of Fields of View

Field of view is the area seen through a microscope. Is the field of view under low power greater than under high power, or are they the same? This exercise will help you answer this question.

1. Make a **hypothesis** to describe how the field of view under low power compares with that under high power. Write your hypothesis in the space provided.

2. Move the slide to a less crowded area of starch grains near the outer edge of the coverslip.

3. Examine the stained starch grains under low power. Count and record the number of grains observed under low power.

_____

4. Without moving the slide, examine the stained starch grains under high power.

a. Count and record the number of grains observed under high power.

_____

b. How does the number of grains observed under low power compare with the number under high power?

_____

When using low power, the total area of your field of view is *greater* than when using high power. The diameter of your low-power field is usually 4 times greater than that of your high-power field. For example, if low power has a magnification of 10× and high power has 40×, divide 40 by 10. Your answer, 4, shows the difference in diameter of these two lenses. Low power has a diameter that is 4 times greater than that observed under high power. Do not confuse diameter observed with total magnification. As magnification increases, diameter observed decreases.

c. How many times greater is the diameter of the low-power field of your microscope than the diameter of the high-power field? Remember that

$$\frac{\text{high-power objective}}{\text{low-power objective}} = \begin{array}{c}\text{number of times low-power} \\ \text{diameter is greater than} \\ \text{high-power diameter}\end{array}$$

_____

d. Your answers to questions 4a and 4b should show about 16 times more starch grains viewed under low power as compared with high power. Explain. (Hint: The field of view is a circle. The area of a circle is equal to $\pi(\frac{1}{2}d)^2$, where d is the diameter of the circle.)

_____

5. Be sure to wash your hands after cleanup is complete.

> **HYPOTHESIS**

_____

_____

_____

_____

Lab 7-2

How Can a Microscope Be Used in the Laboratory?

DATA AND OBSERVATIONS

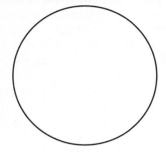

Unstained

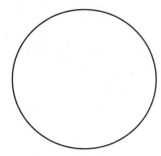

Stained

ANALYSIS

Identify each of the following statements as true or false.

_____ **1.** Objects viewed under the microscope appear upside down.

_____ **2.** When moving the slide toward the left, objects viewed through the microscope will move toward the left.

_____ **3.** The diaphragm is used to adjust the amount of light entering the microscope.

_____ **4.** All objects in different depths appear in focus at the same time while using high power.

_____ **5.** Stains are used to help make clear objects appear lighter under the microscope.

_____ **6.** Low power shows more area than high power.

_____ **7.** High power shows more detail than low power.

_____ **8.** Observers see about 10 times greater diameter under low power than under high power.

_____ **9.** Your depth of field under high power is less than under low power.

CHECKING YOUR HYPOTHESIS

Was your **hypothesis** supported by your data? Why or why not?

_____

_____

FURTHER INVESTIGATIONS

1. Practice using the microscope by inspecting insects that are available locally. Examine the small hairs on the bodies of the insects as well as the eyes, wings, and antennae. Note the effects of changing the diaphragm opening and objective lenses and of moving the slide around.

2. Discuss with your teacher how other stains and methods for staining specific types of plant tissues are used. Practice staining wet mounts from various common garden vegetables and fruits.

EXPLORATION

Normal and Plasmolyzed Cells

Diffusion of water molecules across a cell's plasma membrane from an area of high water concentration to an area of low water concentration is called osmosis. This movement of water may be harmful to cells. If too much water is lost from the cell, the plasma membrane and the cell contents shrink. This is called plasmolysis. Plasmolysis may lead to death of the cell. Most cells live in an environment where movement of water in and out of the cell is about equal. Therefore, there are no harmful effects to the cell.

OBJECTIVES

- Prepare a wet mount of an *Elodea* leaf.
- Observe plasmolysis by adding salt solution to the wet mount.
- Observe the reversal of plasmolysis by diluting the salt solution in the wet mount.
- Compare and diagram normal plant cells in tap water with plasmolyzed plant cells in salt solution.

MATERIALS

compound light microscope
microscope slide
paper towel
coverslip
Elodea (water plant)

droppers (2)
tap water
6% salt solution
forceps
laboratory apron
goggles

PROCEDURE

1. Prepare a wet mount of an *Elodea* leaf as follows. Use Figure 1 as a guide.

2. Use a dropper to place one or two drops of tap water on a microscope slide.

3. Place one leaf, taken from the top whorl of leaves on a sprig of *Elodea*, in the drop of water. Cover the leaf with a coverslip.

4. Observe the leaf under both low and high powers. Note the location of chloroplasts in relation to the cell wall.

5. Diagram a normal cell in the space provided in Data and Observations. Label the cell wall, plasma membrane, and chloroplasts.

6. Use a clean dropper to add a drop of 6% salt solution along one edge of the coverslip. Place a piece of paper towel along the opposite edge of the coverslip. The tap water will soak into the paper towel, drawing the salt solution under the coverslip.

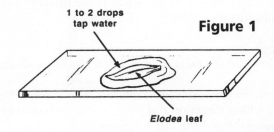

1 to 2 drops tap water

Figure 1

Elodea leaf

7. Observe the leaf under both low and high powers. Again note the location of the chloroplasts in relation to the cell wall. If the plasma membrane and the cell contents have shrunk away from the cell wall, the cell has plasmolyzed.

8. Diagram a plasmolyzed cell in the space provided in Data and Observations. Label the cell wall, plasma membrane, and chloroplasts.

9. Add a drop of tap water to the wet mount, following the procedure in step 6. Observe the appearance of the cells.

Normal and Plasmolyzed Cells

DATA AND OBSERVATIONS

Normal plant cell

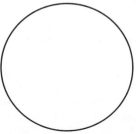

Plasmolyzed plant cell

ANALYSIS

1. Describe the location of chloroplasts in a normal plant cell (in tap water).

_____

2. Describe the location of chloroplasts in a plasmolyzed plant cell (in salt solution).

_____

3. How did the cells change when salt solution was added to the wet mount?

_____

_____

4. In which direction did water move through the plasma membrane of the cells when salt solution was added?

_____

5. How did the cells change when tap water diluted the salt solution?

_____

6. In which direction did water move through the plasma membrane of the cells when tap water diluted the salt solution?

_____

7. Describe the process of plasmolysis.

_____

_____

_____

FURTHER EXPLORATIONS

1. Repeat this Exploration using cells from the epidermis of a purple onion bulb. Compare the results.

2. Repeat this Exploration using 6% glucose solution in place of the salt solution. Compare your results. Explain why the results are different from those in this Exploration.

Copyright © Glencoe/McGraw-Hill, a division of The McGraw-Hill Companies, Inc.

Lab 8-2

INVESTIGATION

How Does the Environment Affect Mitosis?

Mitosis is the division of the nucleus of eukaryotic cells followed by the division of the cytoplasm. If the division proceeds correctly, it produces two cells that are genetically identical to the original cell. Mitosis is responsible for the growth of an organism from a fertilized egg to its final size and is necessary for the repair and replacement of tissue. Anything that influences mitosis can impact the genetic continuity of cells and the health of organisms.

How do environmental factors affect the rate and quality of mitotic division? Scientists are perhaps most keenly interested in this question from the perspective of disease, specifically, the uncontrolled division of cells known as cancer. This investigation will allow you to make a simplified study of the relationship between the environment and mitosis. First, you will observe onion bulbs grow roots by mitotic division. Then you will test the rate of growth of the same onion roots when exposed to an environmental chemical, caffeine in the form of coffee.

OBJECTIVES

- Prepare squashes of onion root tips to observe mitosis.

- Make a hypothesis to describe the effect of caffeine on mitosis.

- Compare growth of onion roots in water and in caffeine.

MATERIALS

onion bulbs (4)
toothpicks (16)
150-mL glass jars (4)
concentrations of caf-
    feine (coffee): 0.1%,
    0.3%, 0.5%
metric ruler
wax marking pencil

scalpel
paper towels
distilled water
microscope slides (4)
coverslips (4)
Feulgen stain
methanol-acetic
    acid fixative

3% hydrochloric acid
45% acetic acid in a
    dropper bottle
forceps
microscope
25-mL graduated
    cylinders (2)
test tubes (8)
test-tube holder

test-tube rack
thermometer
hot plate
water bath
clock or watch
thermal mitts (1 pair)
laboratory apron
goggles

PROCEDURE

Part A. Comparing Rates of Growth

1. Put on a laboratory apron and goggles. Label the small glass jars A, B, C, and D.

2. Insert a toothpick into opposite sides of each onion bulb so that each bulb can be balanced over the mouth of a jar, as shown in Figure 1. Then pour water into each jar until just the root area of the bulb is immersed. Wash your hands thoroughly.

3. Examine the bulbs each day. In Table 1, record the number of roots that emerge from each bulb and the average of their lengths.

Figure 1

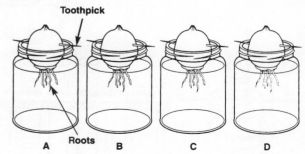

How Does the Environment Affect Mitosis?

PROCEDURE   continued

4. When the roots have grown to 1 cm in length, pour the water out of jars B, C, and D. Your teacher will provide you with caffeine solutions of three different concentrations. Fill jar B with the 0.1% solution, jar C with the 0.3% solution, and jar D with the 0.5% solution. Once again, balance the bulbs over the mouth of jars B, C, and D so that the roots are immersed.

5. Measure the roots for 3 more days, each time recording the average length of the roots for each of the treatments (that is, water and the three concentrations of caffeine) in Table 2.

Part B. Comparing Phases of Mitosis

Note: READ ALL STEPS BEFORE YOU START.

1. Label 4 test tubes A, B, C, and D to correspond to the treatments to which the onion bulbs are being subjected. Then pour 5 mL of methanol-acetic acid fixative into each of the tubes.

2. Set up and begin heating the water bath to 60°C.

3. Use the scalpel to remove all of the roots from each of the onion bulbs. **CAUTION:** *Use the scalpel with care. Cut away from your fingers.* Then use the scalpel to cut a 3 mm piece from the *tips* of each root. Immediately place these tips from onion bulbs treated in A, B, C, and D jars into the corresponding test tubes containing the methanol-acetic acid fixative.

4. Use the test-tube holder to place test tubes A–D into the water bath at 60°C for 15 minutes.

5. Carefully pour the fixative from each tube into a labeled container to be disposed of by the teacher. Transfer the root tips from each tube to four new test tubes labeled A–D.

6. Pour 5 mL of 3% hydrochloric acid into each of the new test tubes in order to prepare the DNA for staining. **CAUTION:** *Hydrochloric acid is a strong acid and causes burns. Avoid contact with skin or eyes. Flush with water immediately if contact occurs and call the teacher.* Place the test tubes into the water bath at 60°C for 10 minutes.

7. Carefully pour the acid into a labeled empty beaker that the teacher has set aside for the acid. Add enough drops of Feulgen stain into each test tube to cover the roots. **CAUTION:** *The stain can discolor your clothes and skin. Use it with care.* Let the tissues sit in the stain for 15 minutes.

8. From tube A, remove one root tip with a pair of forceps. Place the root tip in the center of a labeled slide. Add one or two drops of acetic acid. **CAUTION:** *If acetic acid is spilled, flush with water immediately and call the teacher.* Then place a coverslip over the specimen.

9. Place the slide on a paper towel cushion and cover the slide and coverslip with a piece of paper towel. Push down onto the coverslip with the eraser of a pencil. This is called a squash. *Do not press too hard or you will break the coverslip.*

10. Repeat steps 8 and 9 for treatments B, C, and D.

11. Make a **hypothesis** to describe the effect of caffeine on the stages of mitotic division. Write your hypothesis in the space provided.

Figure 2

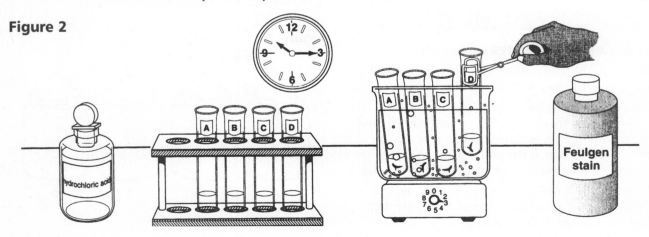

Copyright © Glencoe/McGraw-Hill, a division of The McGraw-Hill Companies, Inc.

How Does the Environment Affect Mitosis?

12. Look at your slides under the microscope at low and high powers for cells undergoing mitosis. The cells will not be as neatly arranged as they would be on prepared slides. Examine the size, shape, and position of chromosomes in each treatment in order to help you identify phases of mitosis. In comparing treatments, do you notice differences in the number of cells in each phase? In the stronger caffeine solutions, do the chromosomes in any particular phase seem especially distinct? Count and record in Table 3 the number of cells in each phase of mitosis.

13. On a sheet of paper, sketch the stages of mitosis observed from roots in each treatment.

HYPOTHESIS

_____

_____

DATA AND OBSERVATIONS

Table 1

	Number of Roots and Average Length in Water							
	Bulb A		Bulb B		Bulb C		Bulb D	
Day	Number	Avg. Length	Number	Avg. Length	Number	Avg. Length	Number	Avg. Length
1								
2								
3								

Table 2

	Number of Roots and Average Length							
	Bulb A (water)		Bulb B (0.1%)		Bulb C (0.3%)		Bulb D (0.5%)	
Day	Number	Avg. Length	Number	Avg. Length	Number	Avg. Length	Number	Avg. Length
1								
2								
3								

Table 3

	Number of Mitotic Phases in Each Treatment					
Treatment	Interphase	Prophase	Metaphase	Anaphase	Telophase	Cytokinesis
Bulb A						
Bulb B						
Bulb C						
Bulb D						

How Does the Environment Affect Mitosis?

Lab 8-2

ANALYSIS

1. Identify the control and variable for the experiment.

_____

_____

2. Study Tables 1 and 2. Compare the rate of growth of the roots immersed in water with the rate of root growth in the various concentrations of caffeine.

_____

_____

3. Describe any differences in the number of cells in each mitotic phase among the four squashes.

_____

_____

_____

4. How do your observations about mitotic phases in Part B relate to your observations about rate of root growth in Part A?

_____

_____

_____

5. What are some conditions or factors in the environment that might have an effect upon the rate or quality of mitotic division?

_____

_____

CHECKING YOUR HYPOTHESIS

Was your **hypothesis** supported by your data? Why or why not?

_____

_____

FURTHER INVESTIGATIONS

1. Design an experiment to test the effects of environmental chemicals on the growth and development of eggs of *Ascaris*, a parasitic roundworm.

2. Design an experiment to test the effect of ultraviolet radiation upon the mitotic division of onion, bean, and *Ascaris* cells.

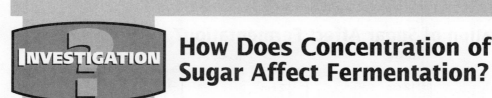

How Does Concentration of Sugar Affect Fermentation?

Yeast are unicellular fungi that obtain their energy by fermenting organic materials. Alcoholic fermentation is a process by which yeast breaks down sugars to produce ethyl alcohol and carbon dioxide. The progress of fermentation can be monitored by examining either ethyl alcohol or carbon dioxide production. Ethyl alcohol can be detected by its distinct odor. Carbon dioxide can be detected by a color change in bromothymol blue.

OBJECTIVES

- Make a hypothesis that describes the effect an increase in sugar concentration will have on yeast respiration.

- Prepare four different growth environments for yeast and measure the rate of respiration in each.
- Draw a graph that compares the rates of respiration for yeast in the four environments.

MATERIALS

colored pencils (4 colors)
molasses solutions (10%, 20%, 50%)
graduated fermentation tubes (4)
bromothymol blue solution
10-mL graduated cylinder
graph paper

wax marking pencil
distilled water
yeast suspension
stopwatch
metric ruler
laboratory apron
goggles

PROCEDURE

1. Make a **hypothesis** to predict how increasing the concentration of sugars will affect the rate of respiration in yeast. Write your hypothesis in the space provided.

2. Label the fermentation tubes "1," "2," "3," and "4," respectively.

3. Add 1 mL of yeast suspension and 1 mL of bromothymol blue solution to each tube.

4. Add enough distilled water to fermentation tube 1 to bring the fluid level in the closed part of the tube to within 2 cm of the top as shown in Figure 1. Mark with your wax marking pencil the level of fluid in the closed end of the fermentation tube.

Molasses is a syrup that contains a variety of sugars. Your teacher will provide you with molasses solutions of three different concentrations. Use these solutions to carry out steps 4, 5, and 6.

5. Add enough 10% molasses solution to fermentation tube 2 to bring the fluid level in the closed part of the tube to within 2 cm of the top. Mark the fluid level with the wax marking pencil.

The fluid level in the closed end of the fermentation tube will be used to measure the amount of gas produced. As gas accumulates, the fluid level in this part of the tube will drop. You will be monitoring how much the fluid level drops over time.

6. Add enough 20% molasses solution to fermentation tube 3 to bring the fluid level in the closed part of the tube to within 2 cm of the top. Mark the level of the fluid in the closed end of the tube as in step 4.

7. Repeat step 5 using the 50% molasses solution in fermentation tube 4.

8. Observe the fermentation tubes and complete the row marked "0 minutes" in Table 1 for each

How Does Concentration of Sugar Affect Fermentation?

Lab 9-1

Figure 1

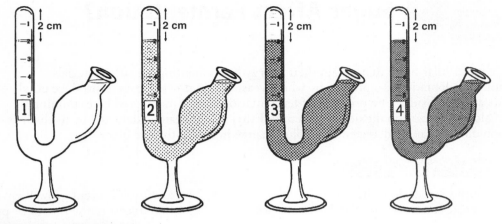

Copyright © Glencoe/McGraw-Hill, a division of The McGraw-Hill Companies, Inc.

PROCEDURE continued

of the four tubes. Begin timing the experiment with a stopwatch.

9. After 10 minutes have elapsed, observe the tubes and complete the row in the table marked "10 minutes" for each of the four tubes.

10. Repeat step 9 three more times at 20 minutes, 30 minutes, and 40 minutes after the start of the experiment. Complete the appropriate rows in the table for each of the four fermentation tubes.

11. Place your fermentation tubes in a protected place.

12. At the next class meeting, observe the tubes and complete Table 1.

13. Clean and return experiment materials to their appropriate places.

14. Construct a graph from the data in Table 1. Plot the amount of gas produced over time. The vertical axis should represent the amount of gas produced (in mL), and the horizontal axis should represent the time elapsed (in minutes). Use a different colored pencil to draw a line connecting the data points for each fermentation tube.

HYPOTHESIS

_____

_____

DATA AND OBSERVATIONS

Table 1

Testing for Respiration Rate				
	Time elapsed	Color of solution	Odor of ethyl alcohol (none, slight, strong)	Gas production (volume in mL)
Tube 1	0 min			
	10 min			
	20 min			
	30 min			
	40 min			
	24 hr			

How Does Concentration of Sugar Affect Fermentation?

Table 1 (continued)

		Testing for Respiration Rate		
	Time elapsed	Color of solution	Odor of ethyl alcohol (none, slight, strong)	Gas production (volume in mL)
Tube 2	0 min			
	10 min			
	20 min			
	30 min			
	40 min			
	24 hr			
Tube 3	0 min			
	10 min			
	20 min			
	30 min			
	40 min			
	24 hr			
Tube 4	0 min			
	10 min			
	20 min			
	30 min			
	40 min			
	24 hr			

ANALYSIS

1. In which fermentation tubes did the fluid color change from blue to yellow? _____

What was the relationship between sugar concentration and the time when the fluid color changed?

_____

_____

2. In which tubes did you detect the odor of ethyl alcohol? _____

In which tube did the odor become noticeable first? _____

How Does Concentration of Sugar Affect Fermentation?

ANALYSIS continued

3. What gas was released in tubes 2, 3, and 4?

_____

How was the presence of this gas demonstrated?

_____

_____

4. What changes took place in tube 1? Explain.

_____

_____

5. In which tube was the most gas produced? Why?

_____

_____

CHECKING YOUR HYPOTHESIS

Was your **hypothesis** supported by your data? Why or why not?

_____

_____

FURTHER INVESTIGATIONS

1. Carry out the same experiment under different conditions. You might keep the molasses concentration constant and vary the amount of yeast suspension in the four tubes.

2. Devise and carry out an experiment to determine the optimum temperature for yeast fermentation.

Observation of Meiosis

Meiosis is a type of cell division that reduces the number of chromosomes to half the number found in body cells. This reduction in chromosome number occurs during gamete production and is necessary in order to maintain a stable number of chromosomes in cells from generation to generation. In flowering plants, meiosis results in the formation of male and female gametes. The male gametes are produced in the anthers of a flower.

OBJECTIVES

- Observe the stages of meiosis in lily anthers.
- Draw and label the stages of meiosis in lily anthers.

MATERIALS

compound light
   microscope
prepared slide of
   a lily anther
drawing paper (optional)

pencil (colored
   pencils if desired)
laboratory apron
goggles

PROCEDURE

1. Place a prepared slide of a lily anther on the microscope under low power.

2. Locate cells in the anther that are undergoing meiosis.

3. Choose a cell in meiosis and identify the stage of meiosis the cell is in by comparing it with the stages in Figure 1.

4. In the space provided in Data and Observations, draw the cell and label it with the name of the appropriate stage of meiosis.

5. Continue to observe, identify, and draw cells for as many different stages of meiosis as you can find.

Figure 1

Prophase I

Late Prophase I

Metaphase I

Anaphase I

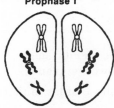
Telophase I

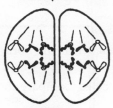
Metaphase II

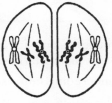
Anaphase II

Telophase II

Observation of Meiosis

DATA AND OBSERVATIONS

ANALYSIS

1. Why do you think lily anthers were chosen for this observation?

_____

_____

2. Which stages of meiosis did you observe most frequently?

_____

_____

3. Describe the chromosomes as they appear in the anther cells.

_____

_____

4. What is the overall function of meiosis in lily anthers?

_____

_____

FURTHER EXPLORATIONS

1. Obtain a textbook from the teacher or the library that identifies and discusses the meiotic stages in humans. Compare and contrast the stages of meiosis in humans with those in flowering plants.

2. Make slides of onion root-tip cells. Compare the cells undergoing mitosis with those of lily anthers undergoing meiosis.

INVESTIGATION

What Phenotypic Ratio Is Seen in a Dihybrid Cross?

The fruit fly, *Drosophila melanogaster*, is one of the most important organisms used by geneticists in studying the mechanisms of inheritance. Fruit flies are used for a number of reasons. Their very short life cycle allows scientists to study several generations in a short time. Several hundred offspring can be produced by each female under ideal conditions. Small vials and simple media enable the easy culture of the insects. There are many mutations available for study in this small animal, which has only eight chromosomes in its diploid cells.

OBJECTIVES

- Learn to care for and raise two generations of fruit flies.

- Construct Punnett squares for two dihybrid crosses.

- Develop hypotheses to describe the phenotypic results of two dihybrid crosses.

- Observe the results of two dihybrid crosses.

MATERIALS

culture vials with medium and foam plugs (2)
culture of vestigial-winged, normal-bodied fruit flies
culture of normal-winged, ebony-bodied fruit flies
vial of alcohol
anesthetic
anesthetic wand

white index card
camel-hair brush
wax marking pencil
stereomicroscope or hand lens
laboratory apron
goggles

PROCEDURE

You will need to learn to identify the sexes of fruit flies. Look at the drawing in Figure 1 for help. The dark, blunt abdomen with dark-colored claspers on the underside identifies the male. Males also have a pair of sex combs on the front pair of legs.

In this Investigation, you will study the inheritance of two traits. Both of these traits are easily observable in Figure 2. The allele for normal wings (*W*) is dominant to the allele for vestigial wings (*w*). Normal long wings that permit flight contrast clearly with the vestigial wings that are short and useless for flying. Normal body color (*B*) is dominant to ebony color (*b*). The normal body color of fruit flies is tan, whereas ebony is a darker brown or black.

The matings in this experiment must begin with unmated (virgin) female flies. This will ensure that

Figure 1

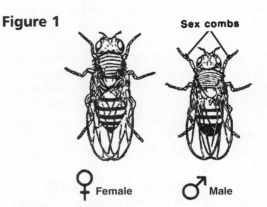

Sex combs

♀ Female   ♂ Male

only the desired males contribute to the offspring of the cross. A female *Drosophila* can store enough sperm from a single mating to fertilize all the eggs she produces in her lifetime. By using virgin females, the parentage of each generation can be controlled.

What Phenotypic Ratio Is Seen in a Dihybrid Cross?

PROCEDURE continued

Figure 2

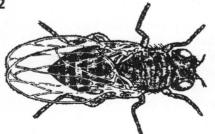

Normal wings (*WW* or *Ww*)
Ebony body color (*bb*)

Vestigial wings (*ww*)
Normal body color (*BB* or *Bb*)

1. Remove all adult fruit flies from both parental cultures when larvae begin to pupate. To remove flies, you must anesthetize them. Tap the vial against a table so that the flies fall to the bottom of the vial. Quickly remove the plug and place a wand containing a few drops of anesthetic into the vial. When the flies are anesthetized, remove the wand from the vial. Remove the adult flies from the vial and place them into a vial of alcohol to kill them. This step will ensure that only virgin females remain in the culture.

2. On the first morning that new adults emerge in the parental cultures, obtain a new culture vial. Label this vial "**P₁ − ♀*wwBB* × ♂*WWbb***" and add your name. This vial will be used to raise the first filial generation.

3. Anesthetize as described above, and collect five virgin females from the vial that contains vestigial-winged, normal bodied flies, and collect five males from the vial that contains normal-winged, ebony-bodied flies. Use a stereomicroscope or hand lens to aid in identification of males and females. Manipulate the flies with the paintbrush. It is easier to examine flies using a piece of white paper or a white index card as a background. Act carefully, but quickly, before the anesthetic wears off.

4. Place the 10 flies into the vial marked "**P₁ − ♀*wwBB* × ♂*WWbb***" and plug the vial with a foam plug. **HINT:** *Always leave the vial on its side until the flies recover from the effects of the anesthetic. This will ensure that they are not injured or killed.* These ten parental flies will mate and produce the first filial generation (F₁). The female

flies will lay eggs, and larvae will appear in 8 to 10 days. When the larvae begin to pupate, remove the parent flies to the vial of alcohol.

5. Construct and complete a Punnett square in the space provided in Data and Observations, showing the possible phenotypes of the F₁ generation.

6. Make a **hypothesis** that predicts the phenotypic ratio of the F₁ generation from this cross. Write your hypothesis in the space provided.

7. Record the numbers and types of parental flies in Table 1.

8. Store your vial of the F₁ generation according to the instructions of your teacher.

9. Label a new culture vial with your name and "**F₂ generation.**"

10. As new F₁ adults emerge over a period of about two weeks, anesthetize and count the numbers and types of flies that appear in this F₁ generation. Record these data in Table 1.

11. Place several of these F₁ adults, as they are anesthetized and counted, into the vial marked "**F₂ generation.**" These F₁ adults will be used to produce the F₂ (second filial) generation. Be sure to include both males and females in the F₂ vial.

12. Construct and complete a Punnett square in the space provided in Data and Observations for this dihybrid cross.

13. Make a **hypothesis** that predicts the phenotypic ratio of the F₂ generation from this dihybrid cross. Write your hypothesis in the space provided.

What Phenotypic Ratio Is Seen in a Dihybrid Cross?

14. When larvae begin to appear in the F_2 vial, remove the adult F_1 flies to the vial of alcohol. Do not add any more F_1 flies to the F_2 vial.

15. As F_2 adults emerge from their pupae, anesthetize and count the different types of flies. Remove flies after they are counted.

16. Record your data for the F_2 generation in Table 1.

17. Record the totals for F_1 and F_2 below Table 1.

18. Dispose of your cultures as directed by your teacher.

HYPOTHESIS 1

_____

_____

HYPOTHESIS 2

_____

_____

DATA AND OBSERVATIONS

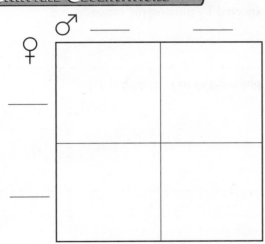

F_1 generation

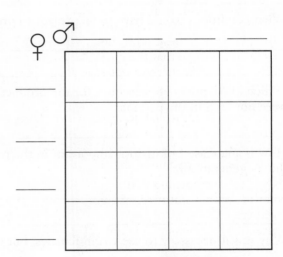

F_2 generation

Table 1

	Results of Dihybrid Crosses			
Generation	Normal-winged normal-bodied	Vestigial-winged normal-bodied	Normal-winged ebony-bodied	Vestigial-winged ebony-bodied
Parental males				
Parental females				
F_1 males				
F_1 females				
F_2 males				
F_2 females				

Total number of F_1 individuals _____   Total number of F_2 individuals _____

What Phenotypic Ratio Is Seen in a Dihybrid Cross?

ANALYSIS

1. What phenotypes were evident in the F_1 generation?

_____

2. What phenotypes were evident in the F_2 generation?

_____

_____

3. According to your Punnett square, what was the expected F_2 phenotypic ratio? What was the actual F_2 phenotypic ratio you observed?

_____

_____

4. What conditions would have to be present to produce the expected F_2 phenotypic ratio?

_____

_____

5. Which of Mendel's laws account for the production of new phenotypes in offspring that were not seen in the parents?

_____

6. How is anaphase I of meiosis important in the production of the observed ratio of traits in the F_2 generation?

_____

_____

7. Why was it necessary to use virgin females, but not virgin males, for this Investigation?

_____

_____

CHECKING YOUR HYPOTHESIS

Were your **hypotheses** supported by your data? Why or why not?

_____

_____

FURTHER INVESTIGATIONS

1. Perform a similar experiment for two generations with fruit flies that have a sex-linked trait, such as white eyes, to determine how the ratio of phenotypes changes.

2. Refer to books on genetics to learn how crosses can be used to determine the relative distances between genes located on the same chromosome.

Lab

11-1

Chromosome Extraction and Analysis

Fruit flies have been used since 1909 as a subject of genetic studies. The cells in the salivary glands of fruit fly larvae have giant polytene chromosomes, which are more than 200 times larger than those in the adult fruit fly. Polytene chromosomes are formed by multiple chromosome replications without any separation of the replicated strands. As a result, the salivary glands of the larva carry many copies of each gene found in the adult fruit fly. When stained, polytene chromosomes have a banded appearance. Scientists have been able to correlate the size, width, and location of the bands with specific genes. As in the cells of the adult fruit fly, the cells from the larva salivary glands have 4 pairs of homologous chromosomes. However, they cannot all be distinguished because the chromosomes in the glands are attached to each other.

OBJECTIVES

- Extract chromosomes from a fruit fly larva.
- Stain and examine the chromosomes under a microscope.

MATERIALS

fruit fly larvae
   in a culture
0.7% saline solution
18% hydrochloric
   acid solution

45% acetic acid
   solution
aceto-orcein stain
droppers (1 for each
   solution)

microscope slides
   and coverslips
paper towels
dissecting probe
forceps, fine-nosed

dissecting microscope
compound light
   microscope
laboratory apron
goggles

PROCEDURE

A female fruit fly can lay up to 500 eggs. Each egg hatches into a larva, which grows and molts twice before forming a hard pupal case. Inside its case, a larva changes into an adult fruit fly.

1. Place a drop of saline solution on a microscope slide. Using forceps, carefully remove a larva from a culture and place it in the drop of saline.

2. Locate the anterior and posterior ends of the larva. The head is the darker area. Also, larvae move in the direction of the anterior end.

3. Use a dissecting microscope to examine the anterior end. Use Figure 2 to help you locate the mouth parts.

4. While viewing the larva under the dissecting microscope, carefully squeeze the larva with forceps about two-thirds of the way back from the mouth parts. (See Figure 2.) This will force

Figure 1

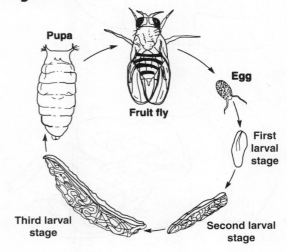

Pupa

Fruit fly

Egg

First
larval
stage

Second larval
stage

Third larval
stage

Copyright © Glencoe/McGraw-Hill, a division of The McGraw-Hill Companies, Inc.

Chromosome Extraction and Analysis

PROCEDURE continued

fluid toward the anterior end and make the head region protrude.

5. With a dissecting probe, pierce the head and gently pull straight forward. If done properly, the salivary glands and other internal structures will emerge in strings. The salivary glands are directly behind the head region. Study Figure 2 so that you do not confuse the salivary glands with intestines or fat glands. The salivary glands will appear transparent and cellular. If you were unsuccessful with the procedure, repeat it with another larva.

6. Remove all parts of the larva except the salivary glands from the slide. Drain the solution from the slide, and carefully dab the slide dry with a paper towel. Do not touch the salivary glands with the toweling.

7. Use a dropper to flood the glands with hydrochloric acid solution for 3 minutes and then drain the slide. **CAUTION:** *If hydrochloric acid is spilled, rinse with water and call the teacher immediately.*

8. Flood the glands with aceto-orcein stain for two minutes. Now tilt the slide and allow the stain to drain to one side. Wipe any excess stain from the slide with a piece of paper toweling, being careful to avoid touching the glands.

9. Look for a dark-red stained area. Then place the slide on a paper towel and gently rinse it with acetic acid, being very careful not to wash away the salivary glands. **CAUTION:** *If acetic acid is spilled, rinse with water and call the teacher immediately.* Drain the excess acid onto a paper towel. Repeat this step until most of the stain has been removed.

10. Carefully lower a coverslip over the salivary glands. Using your thumb, *gently* press the coverslip. Your purpose is to spread the cells into a single layer, at the same time releasing the chromosomes from their nuclei.

11. Observe the slide under low power and high power of a compound light microscope. Sketch the chromosomes in the space provided in Data and Observations.

Figure 2

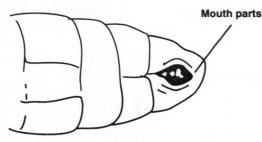

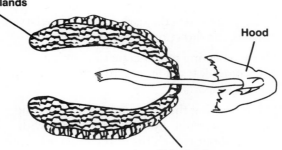

Mouth parts

Salivary glands

Hood

Fat glands

Chromosome Extraction and Analysis

DATA AND OBSERVATIONS

ANALYSIS

1. Describe the structure of the pairs of chromosomes. Why do they have this appearance?

_____

_____

_____

_____

2. The larval stage is a period of enormous growth in the fruit fly's life cycle. What do you think is the advantage of having polytene chromosomes with many sets of genes in the larval stage?

_____

_____

_____

_____

Chromosome Extraction and Analysis

ANALYSIS continued

3. Suppose that you had raised one group of larvae on an enriched culture and another group on a minimal nutrient culture. Would you expect any observable difference in the polytene chromosomes from each group? Why or why not?

_____

_____

_____

_____

4. Compare the chromosomes you observed in this lab with those you observed in mitotic division of onion cells in the BioLab for Chapter 8 of your text. How are they different?

_____

_____

_____

_____

FURTHER EXPLORATIONS

1. Compare the chromosomes in the salivary glands of a variety of insects and describe differences in the numbers and arrangements of the chromosomes.

2. Athletes are often given a chromosome test prior to competing in the Olympics. Cells from inside the mouth are used as a source of chromosomes. Use reference books to determine the procedure used. Include in your research the purpose, the reliability, and the ethics of the test.

EXPLORATION   Isolating Mutants

Electromagnetic radiation constantly bombards us from natural and artificial sources in the environment. These electromagnetic waves have a variety of wavelengths with varying amounts of energy. The shorter the wavelength, the more energetic is the electromagnetic wave. Of the different kinds of electromagnetic radiation, light is the most familiar. Wavelengths of light include those of visible colors as well as the wavelengths of infrared light and ultraviolet (UV) light, which cannot be seen with the human eye. Wavelengths of infrared light are longer than those of visible light; wavelengths of UV light are shorter. UV light is known to damage DNA in cells, which may cause the cells to mutate or die. In this Exploration, you will work with *Escherichia coli* (*E. coli*) bacteria that are normally sensitive to a particular antibiotic. You will expose the bacteria to UV light. Then you will test the exposed bacteria to determine whether any had mutated and became resistant to that antibiotic.

OBJECTIVES

- Expose *E. coli* bacteria to UV light.
- Inoculate and culture bacteria using sterile techniques.

- Isolate *E. coli* mutants that are resistant to an antibiotic.

MATERIALS

disinfectant solution
paper towels
test tube of *E. coli*
    culture in nutrient
    broth
test-tube rack
UV safety goggles

UV light source
clock or watch
incubator (36°–40°C)
sterile cotton swabs (4)
sterile petri dishes
    containing nutrient
    agar (4)

Bunsen burner
striker
beaker
antibiotic disks (16)
forceps
tape

wax marking pencil
metric ruler
laboratory apron
goggles

PROCEDURE

CAUTION: *Work cautiously around open flames. Do not touch eyes, mouth, or any other part of your face while doing this lab. Wear your laboratory apron and goggles at all times. Wash your work surface each day with disinfectant solution, using a paper towel, both before and after doing the lab. Wash your hands with soap and water after working with the bacteria or handling the inoculated agar petri dishes or broth cultures.*

Read all steps before you begin the lab.

Day 1

1. Obtain four petri dishes containing sterile nutrient agar and turn them upside down without opening them. Using a wax pencil,

number the dishes 1 to 4, respectively, in the center of each dish. Then draw perpendicular lines that divide each dish into four quarters, as shown in Figure 1. Turn the dishes right side up.

Figure 1

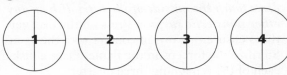

2. Obtain a test tube of *E. coli* broth culture. Remove the stopper from the test tube by

Isolating Mutants

PROCEDURE continued

holding the stopper between your fingers as shown in Figure 2. Do not place the stopper on the table at any time. Holding the test tube near the bottom, slightly tilt the tube and briefly hold the mouth of the test tube in the flame of a Bunsen burner. **CAUTION:** *Pull back loose hair and clothes when using open flames.*

3. Dip a sterile cotton swab into the *E. coli* broth culture. **CAUTION:** *Do not let the swab touch the outside of the test tube or any other surface. Do not let your fingers touch any sterile surface. Use care when working with live bacteria.* Reheat the mouth of the test tube. Replace the stopper on the test tube.

4. Raise the lid of petri dish 1 slightly. Beginning at one end of the dish, quickly rub the cotton swab back and forth across the entire agar surface in a tight S-pattern. To assure coverage of the agar, rotate the dish one-fourth of a turn and repeat steps 3 and 4.

5. Replace the cover of the petri dish. Place the swab in a small beaker of disinfectant.

6. Repeat steps 2–5 for the other three petri dishes.

7. Place petri dishes 3 and 4 on a flat surface. Put on UV safety goggles. Carefully remove the lids from the two petri dishes. Hold the source of UV light 15 cm from the petri dishes and expose the *E. coli* on the dishes to the UV light for one minute. **CAUTION:** *Do not look at the UV light.* Replace the lids on the petri dishes. Do not expose dishes 1 and 2 to UV light.

8. Obtain 16 antibiotic disks from your teacher. Sterilize the ends of the forceps by holding the end in the flame. Slightly raise the lid of one of the petri dishes. Using the forceps, place a disk in the center of one of the four quarters of the agar. **CAUTION:** *Do not let the forceps or disks touch the outside of the petri dish or any other nonsterile surface.* Press on the disk gently to ensure that it is in contact with the agar. Sterilize the forceps in the flame. Place a disk in each of the remaining three quarters. Remember to sterilize the forceps before you

handle each disk. After placing the last antibiotic disk, sterilize the forceps once again.

9. Repeat the process in step 8 with the remaining three petri dishes.

10. Tape the petri dishes closed. Invert the petri dishes and label each with your group name. Incubate the inverted dishes for 24–48 hours.

11. Dispose of the *E. coli* broth culture and cotton swabs as directed by your teacher.

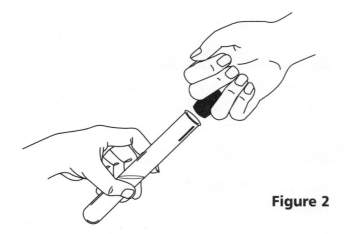

Figure 2

Day 2

1. Remove the petri dishes from the incubator. **CAUTION:** *Do not open the petri dishes at any time.* Examine each dish for bacterial growth. Sketch your results under Data and Observations. Measure and record in Table 1 the diameter (in mm) of the relatively clear area, called the zone of inhibition, around each antibiotic disk. The zone of inhibition is the area where the antibiotic inhibits the growth of antibiotic-sensitive bacteria.

2. For each petri dish, count the number of bacterial colonies growing within each zone of inhibition. Bacterial colonies may be white, yellow, or tan. Fuzzy white growth is likely to be mold. Record your data in Table 1.

3. Dispose of the petri dishes as directed by your teacher. Wash hands thoroughly.

Isolating Mutants

DATA AND OBSERVATIONS

Observation of Incubated Petri Dishes

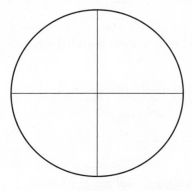

Dish 1
Unexposed *E. coli*

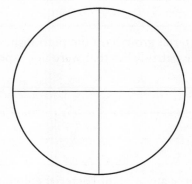

Dish 2
Unexposed *E. coli*

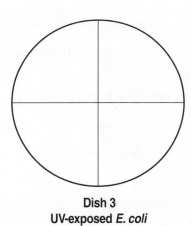

Dish 3
UV-exposed *E. coli*

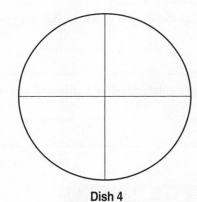

Dish 4
UV-exposed *E. coli*

Table 1

Petri Dish #	Diameters of Zones of Inhibition (mm)				Number of Bacterial Colonies Within Zones of Inhibition			
1. Unexposed *E. coli*								
2. Unexposed *E. coli*								
3. UV-exposed *E. coli*								
4. UV-exposed *E. coli*								

Isolating Mutants

ANALYSIS

1. What was the control in this Exploration?

_____

_____

2. Compare bacterial growth on the petri dishes that were exposed to UV light with bacterial
growth on the petri dishes that were not exposed to UV light.

_____

_____

_____

_____

3. Did you observe any mutant bacterial colonies that are resistant to the antibiotic? Explain.

_____

_____

4. Is there any experimental evidence that suggests that UV light causes mutations?

_____

_____

5. Why were there relatively few antibiotic-resistant colonies on the unexposed petri dishes?

_____

_____

FURTHER EXPLORATIONS

1. Repeat the Exploration using different exposure times to UV light and/or a different type of antibiotic.
Be sure your teacher approves your procedure before you do the experiment.

2. Use library resources to learn about the three kinds of UV light (UVA, UVB, and UVC) and their
effects on organisms. Then read the labels on various sunblock products to find out which of these
types of UV light the products protect against.

EXPLORATION

Determination of Genotypes from Phenotypes in Humans

An organism can be thought of as a large collection of phenotypes. A phenotype is the appearance of a trait and is determined by pairs of genes. The alleles of those genes represent the genotype for the trait. If you were told a large enough number of phenotypic traits that belonged to another person, you would be able to recognize that person.

In this Exploration, you will determine some of your own phenotypic traits. From these, you will be able to determine what your genotypes are for some of the traits. If a trait is dominant and you possess that trait, you will not be able to determine your exact genotype because you could be either homozygous or heterozygous for the gene that controls the trait. However, if a trait is determined by incomplete dominance or codominance, you can tell what your genotype is. Genotypes of recessive traits also can be identified. By comparing your genotypes and phenotypes with those of other people in your class, you will see that you are a unique individual. Given the almost limitless number of gene combinations, it is almost impossible that anyone would have all the same traits as you.

OBJECTIVES

- Determine your phenotype for ten different traits.
- Determine your possible genotypes for the ten different traits.
- Compare your phenotypes and genotypes with those of other students in the class.
- Evaluate your uniqueness as an individual.

PROCEDURE

1. Obtain one piece each of PTC paper and untreated taste paper from the teacher. First, place the untreated paper on your wet tongue to see how it tastes. Then dispose of it in the wastebasket, and place the PTC paper on your wet tongue to see if you can taste phenylthiocarbamide—PTC.

2. PTC is quite bitter and you will notice readily whether or not you have the ability to taste this chemical. If you can taste PTC, enter "taster" in the proper place in the "Your Phenotype" column in Table 1. If you cannot taste the chemical, enter "nontaster" in the table. Discard the taste paper in the wastebasket.

3. Now that you have determined your phenotype, enter in the column marked "Your Possible Genotypes" what your genotype could be. Tasters are either *TT* or *Tt*. Nontasters are *tt*.

MATERIALS

PTC taste paper
untreated taste paper
mirror
laboratory apron
goggles

4. For each of the following traits, observe and record your phenotype in the table. Then record your possible genotypes.

a. hairline—The widow's peak hairline comes to a point in the center of the forehead (*WW* or *Ww*). Individuals that lack the trait are *ww*.

Figure 1

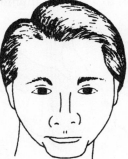

Determination of Genotypes from Phenotypes in Humans

PROCEDURE continued

b. eye shape—Almond-shaped eyes (*AA* or *Aa*) are dominant to round eyes (*aa*).

Figure 2

c. eyelash length—Long eyelashes (*EE* or *Ee*) are dominant to short eyelashes (*ee*).

Figure 3

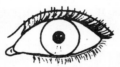

d. tongue rolling—The ability to roll the tongue (*CC* or *Cc*) is dominant to the lack of this ability (*cc*).

Figure 4

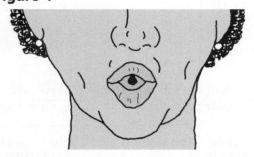

e. thumb—A thumb tip that bends backward more than 30 degrees (hitchhiker's thumb) is dominant (*BB* or *Bb*) to a straight thumb (*bb*).

Figure 5

f. lip thickness—Thick lips (*LL* or *Ll*) are dominant to thin lips (*ll*).

Figure 6

g. hair texture—Curly hair (*HH*) is incompletely dominant to straight hair (*SS*). Wavy hair is *HS*.

Figure 7

h. inter-eye distance—The distance between the eyes is an example of incomplete dominance. Close-set eyes are *DD*, eyes set far apart are *FF*, and medium-set eyes are *DF*.

Figure 8

i. lip protrusion—Protruding lips (*PP*) are incompletely dominant to nonprotruding lips (*NN*). Slightly protruding lips are *PN*.

Figure 9

Determination of Genotypes from Phenotypes in Humans

DATA AND OBSERVATIONS

Table 1

Human Phenotypes and Genotypes				
	Traits		Your phenotype	Your possible genotypes
	Dominant	Recessive		
PTC taste	Taster	Nontaster		
Hairline	Widow's peak	Straight line		
Eye shape	Almond	Round		
Eyelash length	Long	Short		
Tongue rolling	Can roll	Unable to roll		
Thumb	Hitchhiker's thumb	Straight thumb		
Lip thickness	Thick	Thin		
Hair texture	Curly	Wavy	Straight		
Inter-eye distance	Close together	Medium distance	Far apart		
Lip protrusion	Protruding	Slightly protruding	Not protruding		

ANALYSIS

1. Which traits do you have that are dominant?

_____

_____

2. Which traits do you have that are recessive?

_____

_____

3. Which of your traits are governed by incomplete dominance?

_____

_____

4. Which of your traits do you share with one or more of your classmates?

_____

_____

Determination of Genotypes from Phenotypes in Humans

**Lab
12-1**

> **ANALYSIS continued**

5. Which of your traits are unique to you?

_____

_____

6. If you and a particular classmate shared all of the same traits examined in this Exploration, what traits could you describe to prove your uniqueness?

_____

_____

7. What determines your traits?

_____

_____

8. How can a person's genotype for a trait be determined from his or her phenotype for the trait?

_____

_____

_____

_____

9. Why was untreated paper used in the PTC taste test?

_____

_____

> **FURTHER EXPLORATIONS**

1. Books on human genetics from the library or the teacher will discuss many other human traits. Identify some other traits that you or your classmates have and try to determine the genotypes that cause them.

2. Calculate the percentage of the class that has each phenotype and compare these figures with national averages. Suggest reasons why your class might differ from the national percentages of some phenotypes.

Lab

12-2

INVESTIGATION

How Can Karyotype Analysis Detect Genetic Disorders?

A karyotype is a picture in which the chromosomes of a cell have been stained so that the banding pattern of the chromosomes is visible. Cells in metaphase of cell division are stained to show distinct parts of the chromosomes. The cells are then photographed through the microscope, and the photograph is enlarged. The chromosomes are cut from the photograph and arranged in pairs according to size, arm length, centromere position, and banding patterns. Karyotypes have become increasingly important to genetic counselors as disorders and diseases have been traced to specific visible abnormalities of the chromosomes.

OBJECTIVES

- Construct karyotypes from the metaphase chromosomes of six fictitious insects.
- Analyze the karyotypes for chromosome abnormalities.
- Identify the genetic disorders of the insects by using their karyotypes.
- Hypothesize how karyotype analysis can be used to detect genetic disorders.

MATERIALS

photocopies of metaphase chromosomes from
    six fictitious insects (2 pages)
scissors
glue
goggles

PROCEDURE

For this Investigation, assume that a new species of insect has been discovered. The insect has three pairs of very large chromosomes. Researchers have been able to trace four genetic disorders to specific chromosomal abnormalities in this insect. Study the karyotypes and phenotypes of normal male and female insects as illustrated in Figure 1.

Figure 1

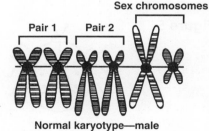

Normal karyotype—male

Normal phenotypic male

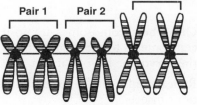

Normal karyotype—female

Normal phenotypic female

How Can Karyotype Analysis Detect Genetic Disorders?

Figure 2

Size reduction disorder
Monosomy of pair 3

PROCEDURE continued

Note that the normal male insect has a pair of sex chromosomes similar to those of the human male, one large and one small. In the same way, the female has a pair of sex chromosomes similar to those of the human female, both large. These sex chromosomes make up chromosome pair 3.

The disorder known as size reduction disorder appears when there is a monosomy of the sex-chromosome pair. A single large sex chromosome produces a small female insect. A single small sex chromosome produces a small male insect. This disorder is shown in Figure 2.

Clear wing disorder, as shown in Figure 3, appears to result from trisomy of the chromosomes of pair 2. The extra chromosome of the second pair produces sterile insects that lack coloring in their wings. Since sterility always results, the clear wing disorder is not passed on to progeny.

A duplication of a portion of a chromosome from pair 1 produces an insect with a double head. This duplication also produces banding on the wings and additional body segments. See Figure 4.

The deletion of a short segment of the large sex chromosome results in a loss of body segmentation and a reduction of body size. This disorder is shown in Figure 5.

1. Obtain copies of the metaphase chromosomes of six insects from the teacher.

2. Write a **hypothesis** that describes how karyotype analysis can be used to detect the presence of a genetic disorder. Write your hypothesis in the space provided.

3. Cut out the chromosomes for insect 1 from the photocopy and place them along the line for insect 1 in Data and Observations. Arrange similar chromosomes together as shown in the normal karyotypes in Figure 1. Match up similar chromosomes by comparing chromosome size, length of the arms of each chromosome, centromere position, and banding patterns. Be sure to line up chromosomes that resemble the first pair of the normal karyotype above the number 1, those that resemble the second pair above the number 2, and those that resemble the third pair (sex chromosomes) above the number 3.

Figure 3

Clear wing disorder
Trisomy of pair 2

Figure 4

Duplication disorder
Duplication of a portion of a chromosome in pair 1

Figure 5

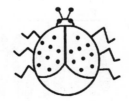

Unsegmented disorder
Deletion of segment of the large chromosome in pair 3

4. Once the chromosomes are positioned, glue their centromeres to the straight line. This represents the karyotype for one insect.

5. Repeat steps 3 and 4 for each of the other fictitious insects.

How Can Karyotype Analysis Detect Genetic Disorders?

6. Compare your karyotypes with the karyotypes of the normal insects and with the descriptions of the genetic disorders.

7. Complete the Analysis for this Investigation.

HYPOTHESIS

_____

_____

_____

DATA AND OBSERVATIONS

Insect 1

_____

1   2   3

Insect 3

_____

1   2   3

Insect 5

_____

1   2   3

Insect 2

_____

1   2   3

Insect 4

_____

1   2   3

Insect 6

_____

1   2   3

How Can Karyotype Analysis Detect Genetic Disorders?

ANALYSIS

1. Identify the sex, chromosome error, and genetic disorder for each of the fictitious insects.

	Sex	Chromosome error	Genetic disorder
Insect 1			
Insect 2			
Insect 3			
Insect 4			
Insect 5			
Insect 6			

2. Which type of chromosome abnormality is the most difficult to detect by means of a karyotype?

The easiest? _____

Why? _____

3. How can duplication of a chromosome of the first pair produce a double head and, at the same time, affect the wing pigmentation and body segmentation?

4. What kind of information is required in using karyotypes to detect genetic disorders in real organisms?

CHECKING YOUR HYPOTHESIS

Was your **hypothesis** supported by your observations? Why or why not?

FURTHER INVESTIGATIONS

1. Books on human genetics will have discussions of chromosome abnormalities and karyotype analysis. Borrow a book from the teacher or the library and investigate some of the common abnormal karyotypes in humans.

2. Obtain additional copies of metaphase chromosomes from the six fictitious insects. Alter the chromosomes and construct new karyotypes using the chromosomes. Imagine how the resulting insects might look. Draw pictures of the new fictitious insects.

Making Test Crosses

EXPLORATION

How do breeders know the genotypes of the animals they are breeding? A test cross is one method of determining the genotype of an organism. A test cross is the mating of an individual of unknown genotype with an individual of known genotype, usually homozygous recessive. From the phenotypes of the offspring, breeders can determine the genotype of the unknown parent. In this exploration, you will be working with the fruit fly, *Drosophila melanogaster*. This species has been used extensively in the study of genetics and inheritance. These fruit flies are ideal for research because they are easily handled, they produce many offspring in a short time, they have few chromosomes, and they have many mutations that can be observed.

OBJECTIVES

• Learn to care for and raise two generations of fruit flies.

• Perform two test crosses with fruit flies.

• Observe the phenotypic results of the two test crosses.

• Infer the genotypes of the parental fruit flies and their offspring.

• Construct Punnett squares for two test crosses.

MATERIALS

culture vials with
    medium and foam
    plugs (2)
culture of vestigial-
    winged fruit flies

culture of normal-
    winged fruit flies
vial of alcohol
anesthetic

anesthetic wand
white index card
fine-tipped paintbrush
wax marking pencil

stereomicroscope or
    hand lens
laboratory apron
goggles

PROCEDURE

Part A. Test Cross 1

Use Figure 1 to identify the sexes of fruit flies. The dark, blunt abdomen with dark-colored claspers on the underside identifies the male. Males also have a pair of sex combs on the front pair of legs.

In this Exploration, you will perform two test crosses to determine the genotypes of two cultures of fruit flies that have normal wings. Use Figure 2 to identify a recessive trait, vestigial wings. Normal long wings (W) are dominant to vestigial short wings (w). Normal long wings enable fruit flies to fly, whereas fruit flies with vestigial short wings are unable to do so.

Figure 1

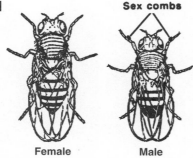

Sex combs

Female   Male

Figure 2

Normal wings (W)   Vestigial wings (w)

Making Test Crosses

PROCEDURE continued

1. When larvae in the normal-winged culture and the vestigial-winged culture begin to form pupae, remove all adult fruit flies from both parental cultures by anesthetizing them. To anesthetize the flies, tap each vial against a table so that the flies fall to the bottom of the vial. Quickly remove the plug and place a wand containing a few drops of anesthetic into the vial. When the flies are anesthetized, remove the wand from the vial. Place the adult flies in a vial of alcohol to kill them. This step will ensure that only virgin (unmated) females remain in the culture. Virgin female flies must be used for these test crosses to ensure that only the chosen males contribute sperm to the offspring of the cross. A female fruit fly can fertilize all the eggs she produces in her lifetime from stored sperm from a single mating. **CAUTION:** *Be careful not to spill alcohol on your clothing or to get it in your eyes. In case of spills, clean up immediately and wash your hands. Do not use alcohol near open flames.*

2. On the first morning that new adults emerge in the parental cultures, obtain a new culture vial. Label this vial "TC_1—female vestigial × male normal" and add your name.

3. Anesthetize and collect five virgin females from the vial of vestigial-winged flies, as well as five males from the vial of normal-winged flies. Use a stereomicroscope or hand lens to aid in identification of males and females. Use a paintbrush to pick up and move the flies. Place a white index card under the flies as a background for easier observation. Act carefully but quickly, before the anesthetic wears off.

4. Place the ten flies into the vial marked "TC_1" and plug the vial with a foam plug. **HINT:** *To ensure that the flies are not harmed, always leave the vial on its side until the flies recover from the effects of the anesthetic.* These ten parental flies will mate, the female flies will lay eggs, and larvae will appear in 8 to 10 days. These offspring are the results of the first test cross (TC_1). When the larvae begin to form pupae, remove the parental flies and kill them by placing them in the vial of alcohol.

5. Record the numbers and types of parental flies in Table 1.

6. Store your TC_1 vial according to your teacher's instructions.

7. As the new TC_1 adults emerge over a period of about two weeks, anesthetize and count the numbers and types of flies that appear in this TC_1 generation. Record these data in Table 1. You may wish to carry out Part B, steps 1 and 2 as you work on this step.

Part B. Test Cross 2

1. Label a new culture vial "TC_2—female vestigial × male TC_1," and add your name.

2. As you anesthetize and count TC_1 adults from Part A, place five male TC_1 adults into the TC_2 vial.

3. Collect five virgin females from the original culture of vestigial-winged fruit flies. Place these in the TC_2 vial with the five males. These ten fruit flies will be used in the second text cross.

4. Record the numbers and types of parental flies in Table 2.

5. When larvae begin to appear in the TC_2 vial, remove the adult flies and kill them in the vial of alcohol. Do not add any more flies to the TC_2 vial.

6. As TC_2 adults emerge from their pupae, anesthetize and count the different types of flies. Remove flies to the vial of alcohol after they have been counted.

7. Record your data for the TC_2 generation in Table 2.

8. Store or dispose of your cultures as directed by your teacher.

Making Test Crosses

DATA AND OBSERVATIONS

Table 1 Test Cross 1

Number of Flies of Each Wing Type		
Generation	Normal-winged	Vestigial-winged
Parental males		
Parental females		
TC_1 males		
TC_1 females		

Table 2 Test Cross 2

Number of Flies of Each Wing Type		
Generation	Normal-winged	Vestigial-winged
TC_1 males		
Vestigial females		
TC_2 males		
TC_2 females		

ANALYSIS

1. What were the genotypes of the male and female parental flies from the original cultures? Explain.

_____

_____

_____

2. What were the phenotypes of the offspring flies in the TC_1 generation?

_____

3. What was the genotype of the offspring flies in the TC_1 generation? Explain.

_____

_____

_____

Making Test Crosses

ANALYSIS continued

4. Use your answers from Analysis questions 1 and 3 to construct a Punnett square for Test Cross 1.

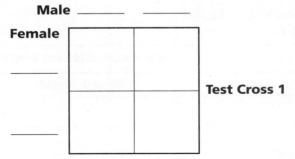

Male _____ _____

Female

_____

_____

Test Cross 1

5. What were the genotypes of the male and female parent flies used in Test Cross 2?

_____

6. What were the phenotypes of the flies in the TC_2 offspring?

_____

_____

7. What were the genotypes of the offspring in the TC_2 vial? Explain.

_____

_____

_____

8. Use your answers from Analysis questions 5 and 7 to construct a Punnett square for Test Cross 2.

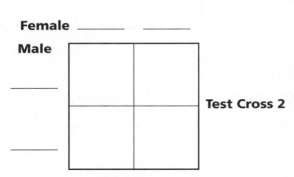

Female _____ _____

Male

_____

_____

Test Cross 2

9. Based on Punnett square 2, what was the expected TC_2 phenotypic ratio? What was the actual TC_2 phenotypic ratio observed?

_____

_____

FURTHER EXPLORATIONS

1. Contact an animal breeder. A listing of breeders near your home can be found on the Internet. Interview the breeder to learn about the methods breeders commonly use to select desired traits in animals. Find out whether the breeder has used test crosses to determine the genotypes of certain animals. Share what you learned with the class.

2. Research a certain breed of dog, cat, or other type of domesticated animal. Write a report that describes the distinctive traits and the history of that breed.

DNA Sequencing

The process of identifying the order of nitrogen bases within a segment of DNA or a gene is called DNA sequencing. It begins with cutting the DNA using restriction enzymes and making many copies of the DNA. The DNA strands are separated, and one end of each strand is labeled with a radioactive probe. The multiple copies of this single-stranded DNA are placed into four test tubes along with the nucleotides and enzyme needed to make new DNA. Each test tube also contains one of four special forms of nitrogen bases A, T, C, and G that stop DNA synthesis at a specific place. As a result, new DNA segments of varying lengths are produced in the tubes. The segments are then sorted by size in a gel electrophoresis. A sheet of X-ray film is laid on top of the gel, and dark bands are formed on the film where the radioactive probes are located. The sequence of the nitrogen bases in the gene being studied can be "read" directly from the film. From this sequence, a scientist can determine the sequence of amino acids in the protein that is coded for by the gene.

OBJECTIVES

- Learn to read a gel electrophoresis.
- Model the process of DNA sequencing, using gel electrophoresis.
- Use an mRNA codon chart to determine the sequence of amino acids coded for by a gene.

MATERIALS

photocopies of strands
   of DNA (4 sheets)
colored pencil
scissors

meter stick
large poster board
   (approximately
   24" × 36")

tape
plastic trays or rectan-
   gular lids, small (4)
wax marking pencil

marker
goggles

PROCEDURE

Part A. Reading a Gel Electrophoresis

1. Figure 1 represents an X ray of the gel from the electrophoresis of segments of a DNA strand. Each letter at the top represents one of the four nitrogen bases in the nucleotides of a DNA molecule. The marks under each of the bases represent segments of DNA that migrated through the gel. (The radioactive probes attached to the segments "burn" these marks into the X-ray film when it is exposed to the gel.) The numbers represent the relative distances traveled by the segments, with "1" being the shortest distance. The shorter the segment, the longer the distance it migrates. The DNA sequence is read *from the bottom of the X ray to the top*, that is, from the

Figure 1

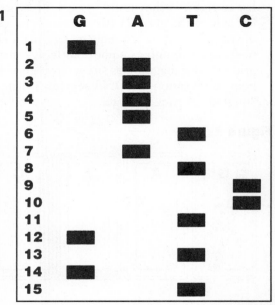

DNA Sequencing

PROCEDURE continued

Figure 2   radioactive probe

ACGGGACTACCATGGGCCTTA

shortest segment of the DNA strand to the longest.

2. "Read" the DNA sequence from the X ray of the gel electrophoresis in Figure 1, and record it in Data and Observations.

Part B. Modeling DNA Sequencing

1. Cut out the 24 photocopied DNA strands as shown in Figure 2. Color the radioactive probe at the left end of each copy of the strand.

2. Place 6 of the strands into each of the plastic trays. The trays represent test tubes containing different chemical treatments. As shown in Figure 3, label the trays left to right with the letters G, A, T, and C to show the special form of the nitrogen base in each tube.

3. On your poster board, use your marker to draw a diagram of an electrophoresis gel with wells at the top where you will place segments of the DNA strands. Label the wells G, A, T, C, from left to right, as shown in Figure 4. Number the left side of the gel 1-21 (to represent the 21 nitrogen bases you are sequencing), making sure to space the numbers equidistantly. Construct a grid on your diagram, as shown in Figure 4, to help you position the DNA segments that "migrate" through the gel.

4. In "test tube" G, cut diagonally through a G on each strand to stop DNA synthesis. *Make sure to cut the G at a different location on each strand so that you will have segments of DNA of different lengths, each with a radioactive probe attached.* Place the segments in the appropriate well of your gel template. Repeat this procedure with the DNA strands in test tubes T and C, cutting T and C, respectively.

5. In test tube A, cut one nitrogen base A on each of the strands. (In the laboratory, if DNA synthesis stops at the first base after the radioactive label, the DNA segment will not move. Scientists "glue on" a starter DNA sequence to avoid this problem.) Place the segments in their well.

Figure 3

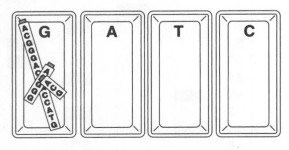

Figure 4

	G	A	T	C
1				
2				
3				
4				
5				
6				
7				
8				
9				
10				
11				
12				
13				
14				
15				
16				
17				
18				
19				
20				
21				

Lab

DNA Sequencing

13-2

6. From well G, move the segments through the gel according to their size. The shortest piece will move the farthest; the longest piece will move the shortest distance. (For example, a segment with one base would move to position 21, a segment with three bases would move to position 19, and so forth.) Do not use the segments without the radioactive label. These segments would move through the gel but would not be visible on the X-ray film. Uncut segments will remain in the wells. Tape each segment in place.

7. Repeat step 6 for the segments from wells A, T, and C.

8. "Read" the DNA sequence from your gel electrophoresis in the same way you read the X ray of the gel electrophoresis in Figure 1. Record your results.

DATA AND OBSERVATIONS

Part A

DNA sequence from gel electrophoresis in Figure 1:

Part B

DNA sequence from your gel electrophoresis:

ANALYSIS

Second Base in Code

		A	G	U	C	
		Lysine	Arginine	Isoleucine	Threonine	A
	A	Lysine	Arginine	Methionine	Threonine	G
		Asparagine	Serine	Isoleucine	Threonine	U
		Asparagine	Serine	Isoleucine	Threonine	C
		Glutamic acid	Glycine	Valine	Alanine	A
First Base in Code	G	Glutamic acid	Glycine	Valine	Alanine	G
		Aspartic acid	Glycine	Valine	Alanine	U
		Aspartic acid	Glycine	Valine	Alanine	C
		STOP	STOP	Leucine	Serine	A
	U	STOP	Trytophan	Leucine	Serine	G
		Tyrosine	Cysteine	Phenylalanine	Serine	U
		Tyrosine	Cysteine	Phenylalanine	Serine	C
		Glutamine	Arginine	Leucine	Proline	A
	C	Glutamine	Arginine	Leucine	Proline	G
		Histidine	Arginine	Leucine	Proline	U
		Histidine	Arginine	Leucine	Proline	C

Third Base in Code

**Lab
13-2**

DNA Sequencing

> **ANALYSIS continued**

1. What is the mRNA sequence that can be transcribed from the DNA sequence in Part A?

_____

_____

2. Refer to the table of mRNA codons on the previous page. Give the amino acid sequence that is coded by the mRNA sequence from question 1.

_____

_____

3. How does the polypeptide coded for by the DNA sequence in Part A differ from most proteins?

_____

_____

4. What is the mRNA sequence that can be transcribed from the DNA sequence in Part B?

_____

_____

5. Refer to the table of mRNA codons again. Give the amino acid sequence that is coded by the mRNA sequence from question 4.

_____

_____

6. How are the strands of "DNA" you studied in this lab different from an actual gene?

_____

_____

> **FURTHER EXPLORATIONS**

1. Research and write an essay on the use of DNA sequencing to identify human genetic disorders, such as cystic fibrosis and hemophilia.

2. Research the topic of genetic markers and design an experiment for identifying markers using gel electrophoresis.

Analyzing Fossil Molds

Brachiopods are solitary, bivalved animals, as shown in Figure 1. All modern brachiopods live in the sea and, based on the fossil record, it is likely that extinct species also were marine organisms. In this lab, you will use two plastic deposits of fossil brachiopods to model the analysis of fossil molds found in sedimentary rock. The deposits will be labeled A and B to distinguish two groups of brachiopods. From measurements of the molds, you will speculate about whether the fossils are from the same or different populations and whether they would have been found in the same layer of sedimentary rock.

OBJECTIVES

- Measure fossil brachiopod molds in two different deposits and record data.
- Prepare a graph of the data from the two deposits.
- Analyze the graph.
- Interpret results relative to geologic time.

MATERIALS

plastic forms of fossil brachiopod deposits, A and B
calipers
metric ruler
colored pencils (2 colors)
graph paper
goggles

PROCEDURE

1. Label one form *A* and the second form *B*. Each form represents a fossil deposit.

2. The length of the brachiopod molds will be used as an indicator of brachiopod size. Using the caliper provided, measure the length of each mold on deposit A. Record each different size in Table 1, as well as the number of molds of that size. If you need more room for measurements, continue the table on a separate sheet of paper.

3. Repeat step 1, using deposit B.

4. Use data from Table 1 to make a line graph or bar graph. The *x*-axis will represent mold length in millimeters; the *y*-axis will represent the number of molds at that length. Use a different colored pencil for each set of molds. For assistance in graphing, refer to *Make and Use Graphs* in the *Skill Handbook* in your textbook.

5. Review the definition of the term *mean*. The *mean* of a set of data is the arithmetic average. It is found by adding the numbers in the data set and dividing it by the number of items in the set.

Figure 1

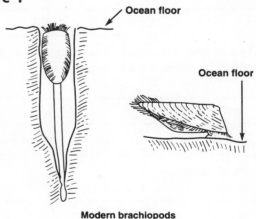

Modern brachiopods

Analyzing Fossil Molds

DATA AND OBSERVATIONS

Table 1

"Deposit A" Brachiopods		"Deposit B" Brachiopods	
Length in mm	Number of molds	Length in mm	Number of molds

ANALYSIS

1. What is the mean size of the molds in deposit A? In deposit B?

_____

2. Describe the shape of the graph for the A molds and for the B molds.

_____

3. Do you think the brachiopods on deposit A represent the same population as the brachiopods on deposit B? Explain your answer.

_____

_____

4. Do you think the molds in deposit A came from the same time period as the molds in deposit B? Explain your answer.

_____

_____

5. Using the concept of relative dating and assuming that the average size of brachiopods increased over time, which deposit of brachiopods would have been found in a more recent layer of rocks?

_____

_____

FURTHER EXPLORATIONS

1. Conduct research to find out why fossils of marine animals, such as brachiopods, are more useful in relative methods of dating sedimentary rocks than are fossils of land animals.

2. There are about 260 living species of brachiopods but more than 30 000 known fossil species. Explore possible reasons for the extinction of so many species.

Making Model Fossils

Fossils are evidence of past life and can provide clues about the characteristics of ancient species. Different types of fossils form in different ways. In this Exploration, you will make models of fossil imprints, molds, and casts. You will use the models to make inferences about real fossils.

OBJECTIVES

- Use common objects to make models of fossil imprints, molds, and casts.
- Use the models to make inferences about real fossils.

MATERIALS

various small objects, such as paper clips, marbles, coins, shells, and leaves
petroleum jelly
modeling clay
wax paper
rolling pin

plaster of paris
cup
water
plastic spoon
goggles
laboratory apron

PROCEDURE

1. Collect a variety of small objects that vary in shape, thickness, and texture. Use a small amount of petroleum jelly to cover the surfaces of the objects.

2. Obtain some modeling clay and begin working it with your hands until it is soft. Divide the clay into as many pieces as you have objects. Flatten each piece of clay onto a sheet of wax paper placed on your lab table.

3. Use your hands or a rolling pin to press each object into the surface of one of the pieces of clay. Remove the objects from the clay, being careful not to gouge any of the clay.

4. Record your observations of the model fossils you made by making drawings in Data and Observations. Decide whether each fossil is an

imprint or a mold. (A mold has more dimension than an imprint.) Under each drawing, identify the object that was used to make the model fossil.

5. Use a spoon and water to make a mixture of plaster of paris in a cup. Follow the directions on the package of plaster of paris. Carefully pour some of the mixture into each of the fossil molds you made. Fill the molds to the top. **NOTE:** *Do not discard any plaster of paris down sink drains.* Allow the mixture to harden overnight in the molds.

6. Carefully remove the plaster casts from the molds. Make drawings of the casts in Data and Observations and identify the object that was used to make each cast.

Copyright © Glencoe/McGraw-Hill, a division of The McGraw-Hill Companies, Inc.

Making Model Fossils

DATA AND OBSERVATIONS

Model Imprints

Model Molds

Model Casts

ANALYSIS

1. Based on your observations, what kinds of remains of living things are more likely to leave imprints rather than molds in rock? Explain.

_____

_____

2. What part of the lab simulated the decay of an organism trapped in sediment?

_____

_____

3. Which of the casts you made were the "best?" Explain.

_____

_____

4. Describe the kinds of remains that would form the most detailed fossil casts.

_____

_____

FURTHER EXPLORATIONS

1. Devise ways to make models of other types of fossils, such as trace fossils, amber- or ice-preserved fossils, and petrified fossils.

2. Fossils can form in many places, however, certain locations produce more fossils than do others. Research fossil formation and the best places for fossils to form. Include the most unusual places as well.

EXPLORATION   Plant Survival

Predator-prey relationships are not confined to the animal kingdom. Enormous numbers of insects and other animals prey on plants, which they damage or destroy in the process of feeding on them. Since plants cannot run away, they have had to evolve a variety of other defense mechanisms to avoid decimation by hungry herbivores. Chemical defenses are common. Nearly all species of plants produce chemicals that are distasteful or dangerous to their potential enemies. Many of these chemicals were probably, at first, metabolic waste products that accumulated in plants. In this Exploration, you will observe the effect of plant chemicals on the behavior of the land snail, which normally consumes a wide variety of plant tissues.

OBJECTIVES

• Prepare a feeding tray containing plant tissues soaked in various chemicals produced by plants.

• Analyze the effects of the different chemicals on the feeding behavior of a land snail.

MATERIALS

land snail
three plant extracts
water
scissors

large leaf lettuce (firm) or Chinese cabbage
graph paper
metric ruler

plastic feeding container (approximately 20 cm in width) with perforated lid

goggles
wax marking pencil
paper towels
laboratory apron

PROCEDURE

1. Cut a piece of graph paper 3.5 cm square. Use it as a template for cutting 4 pieces of lettuce or cabbage. Avoid cutting the leaf along its larger veins. Save the template.

2. Label the bottom of the feeding container with the numbers 1–4 evenly spaced along the width of the container, as shown in Figure 1.

3. Soak one square of the lettuce or cabbage in water for 5 minutes. Remove the square and allow the excess water to drip onto a paper towel. Then place the leaf square on the bottom of the feeding container below "1."

4. Soak and drain each of the remaining leaf squares in a different extract, as described in step 3. Place each square below a number in the feeding container. Record the location of each extract in Table 1.

5. Place the snail in the container. Cover the container with its lid and place it in an area of the room designated by the teacher.

Figure 1

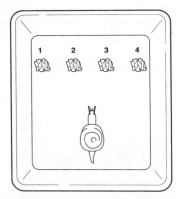

6. After 24 hours, remove the remains of one of the leaves and place it on the square of graph paper that you used as a template. Count and record in Table 1 the number of small squares, or grids, that were consumed from the leaf by the snail. Divide this figure by the number of grids on the square of graph paper to find the percent of the leaf that was consumed. Record the percent in Table 1.

7. Repeat step 6 for the remaining leaves.

Copyright © Glencoe/McGraw-Hill, a division of The McGraw-Hill Companies, Inc.

Plant Survival

DATA AND OBSERVATIONS

Table 1

Type of Extract	Number of Squares Consumed	Percent of Squares Consumed
1. water		
2.		
3.		
4.		

ANALYSIS

1. Which leaves were not eaten by the snail? Which leaves were eaten the most?

_____

_____

_____

2. What is the evolutionary significance of plant chemical defenses?

_____

_____

_____

_____

3. What are some nonchemical plant defenses?

_____

_____

_____

FURTHER EXPLORATIONS

1. Monarch butterfly caterpillars feed on milkweeds, consuming large quantities of alkaloids, which remain in the body of the adult during metamorphosis. The alkaloids make the butterfly distasteful to birds. Design an experiment to test whether monarch butterflies raised on other plants are equally distasteful to birds.

2. Research the coevolution of cabbage butterflies and plants of the mustard family. Discover how these butterflies have overcome the defensive chemicals that are toxic to most insects. Explain the evolutionary advantage of the cabbage butterfly's specialized diet.

Lab
15-2

INVESTIGATION

How Is Camouflage an Adaptive Advantage?

Natural selection can be described as the process by which those organisms best adapted to the environment are more likely to survive and reproduce than are those organisms that are poorly adapted. Organisms have developed many different kinds of adaptations that help them survive in their environments. These include adaptations for finding food, such as keen night vision in nocturnal animals, as well as adaptations for avoiding predators. Some organisms use camouflage as a way to escape predation from other organisms. Camouflage allows them to blend in with the background.

OBJECTIVES

- Use an artificial environment to model the concept of natural selection.
- Hypothesize what will happen if natural selection acts over time on organisms exhibiting camouflage.
- Construct bar graphs to show the results of the Investigation.
- Compare the model of natural selection in the Investigation to real examples of natural selection.

MATERIALS

hole punch
colored paper (1 sheet each of purple, brown, blue, green, tan, black, orange, red, yellow, and white)
plastic film canisters or petri dishes (10)

piece of brightly colored, floral fabric (80 cm × 80 cm)
graph paper (2 sheets)
goggles

PROCEDURE

1. Work in a group of four students.
2. Punch 20 dots from each sheet of colored paper and place each color dot in a different plastic container.
3. Spread out the floral cloth on a flat surface.
4. Spread 10 dots of each color randomly over the cloth. See Figure 1.
5. Select a student to choose dots. That student must look away from the cloth, turn back to it, and then immediately pick up the first dot he or she sees.
6. Repeat step 5 until 10 dots have been picked up. Be sure the student looks away before a selection is made each time.
7. Record the results in Table 1. Return the 10 collected dots to the cloth in a random manner.

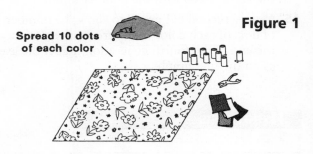

Figure 1

Spread 10 dots of each color

Assume that the dots represent individual organisms that, if allowed, will reproduce more of their own type (color). Also assume that the selection of dots represents predation.

8. Write a **hypothesis** to predict what will happen over time if selected dots are not returned to the cloth and the remaining dots "reproduce." Write your hypothesis in the space provided.

How Is Camouflage an Adaptive Advantage?

PROCEDURE continued

Figure 2

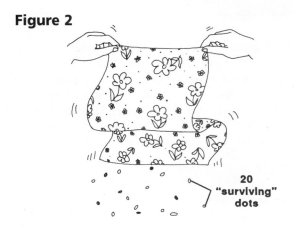

20
"surviving"
dots

9. Each student in the group must, in turn, pick up 20 dots following the method in steps 5 and 6. Place the dots in their original containers. Remember to look away each time before making a selection.

10. After each student has removed 20 dots, shake the remaining 20 dots off the cloth onto the table. See Figure 2.

11. Count and record in Table 2 the number of dots of each color that remains.

12. Give each of the "surviving" dots four "off-spring" of the same color by adding dots from the containers. You may need to punch out more of certain colors. Return all of the dots to the cloth in a random manner. This will bring the total number of dots on the cloth back to 100. See Figure 3.

13. Repeat steps 9–12 three more times. Each repetition represents the survival and reproduction of a single generation. Continue to record the results of each repetition in Table 2.

14. Make a bar graph to show the number of dots of each color that were on the cloth at the beginning of the Investigation. Label the horizontal axis with the names of the 10 colors and the vertical axis with the number of dots.

15. Make a second bar graph to show the number of dots of each color that were on the cloth at the end of the fourth generation. Label the axes as on the first graph.

Figure 3

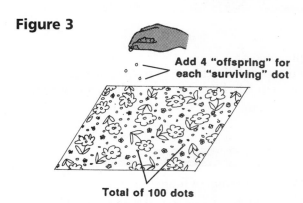

Add 4 "offspring" for
each "surviving" dot

Total of 100 dots

HYPOTHESIS

_____

_____

_____

Lab
15-2

How Is Camouflage an Adaptive Advantage?

DATA AND OBSERVATIONS

Table 1

Selection of Dots	
Color	Number of dots selected
Purple	
Brown	
Blue	
Green	
Tan	
Black	
Orange	
Red	
Yellow	
White	

Table 2

Number of Dots Remaining After Each Generation				
	Number remaining after generation			
Color	1	2	3	4
Purple				
Brown				
Blue				
Green				
Tan				
Black				
Orange				
Red				
Yellow				
White				

ANALYSIS

1. Which colors were picked up from the floral background?

_____

_____

2. Which colors, if any, were not picked up? Why not?

_____

_____

3. If the dots represent food to a predator, what is the advantage of being a color that blends in with the background?

_____

_____

4. Give two examples of real organisms that use camouflage to avoid predation.

_____

_____

How Is Camouflage an Adaptive Advantage?

ANALYSIS continued

5. As the dots on the cloth passed through several generations, what trends in frequency of colors did you observe?

_____

_____

6. How would the outcome of this Investigation have differed if the "predator" was color-blind? Explain.

_____

_____

7. How would the outcome of this Investigation have been affected if dots that were subject to predation (those picked up) tasted bad or were able to harm the predator in some way, such as by stinging it?

_____

_____

_____

8. Describe an example of natural selection that is similar to the model of natural selection in this Investigation.

_____

_____

_____

CHECKING YOUR HYPOTHESIS

Was your **hypothesis** supported by your data? Why or why not?

_____

_____

FURTHER INVESTIGATIONS

1. Certain species of fish are dark-colored on the dorsal side of their bodies and light-colored on the ventral side. Research this adaptation and show how it is an example of camouflage.

2. Design an experiment to show that a chameleon changes its color in response to the color of its background.

EXPLORATION   **Primate Characteristics**

Primates are a group of mammals with large, rounded heads and flattened faces, traits that distinguish them from most other mammals. Their eyes face forward, allowing binocular vision. Their brains are also more complex than those of other animals, allowing complex behaviors. They have five digits on both their hands and their feet. The digits on the hand, and in some species, the feet, are flexible and capable of grasping. Primates can stand upright and walk on two feet. This group includes monkeys, chimpanzees, gorillas, and humans.

OBJECTIVES

- Examine the characteristics of primates.
- Compare the anatomy and capabilities of primates.

MATERIALS

sheet of paper   textbook
pencil   goggles
eraser

PROCEDURE

Part A. Jointed Fingers

Flexible fingers and opposable thumbs are two characteristics of primates.

1. Keep your fingers and your thumb straight (not bending at the joints), and try to pick up the following objects with one hand: a single sheet of paper, a pencil, an eraser, and a textbook. Try to pick up other objects around your classroom that have different shapes and sizes. **CAUTION: *Do not attempt to pick up sharp objects or objects that are capable of breaking.***

2. Try to pick up the same objects once more, but this time bend your fingers.

3. Record your observations in Table 1.

Part B. Comparing Primate Hands

Study the drawings of the hand of a gorilla and a human in Figure 1. Compare the relative sizes and shapes of the thumbs, fingers, and palms. Record your observations in Table 2.

Part C. Binocular Vision

Primates' eyes are located in the front part of their heads, which allows for binocular vision.

1. Close your left eye and look straight ahead. Describe your field of vision—the total area you can see. Record your observations in Table 3.

2. Repeat Step 1, but this time, cover your right eye.

3. Look straight ahead with both eyes open. Describe your field of vision and record your observations in Table 3.

4. Close your eyes, and have a classmate hold up two hands a few feet in front of you, one slightly closer to you than the other. Now open your right eye and try to judge which of your classmate's hands is closer to you. Record your observations in Table 3.

5. Repeat Step 4, but this time open your left eye.

6. Repeat Step 4 one more time with both eyes open.

Part D. Comparing Primate Skeletons

1. Figures 2 and 3 illustrate different aspects of gorilla and human skeletons.

2. Examine Figure 2 and Figure 3. Compare the jaws, skulls, and spinal columns of the gorilla and human. Write your observations in Table 4.

Figure 1   **Gorilla hand**   **Human hand**

Primate Characteristics

Figure 2

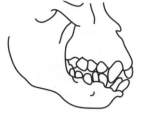
Gorilla jaw

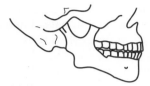
Human jaw

Figure 3

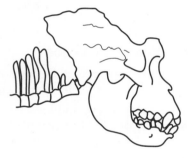
Gorilla skull

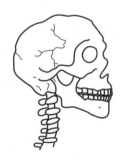
Human skull

DATA AND OBSERVATIONS

Table 1

Object	Using Straight Fingers	Using Bent Fingers
Sheet of paper		
Pencil		
Eraser		
Textbook		

Primate Characteristics

Table 2

	Similarities Between Gorilla and Human Hands	Differences Between Gorilla and Human Hands
Thumb		
Fingers		
Palm		

Table 3

	Field of Vision	Which Hand is Closer?
Using only the right eye		
Using only the left eye		
Using both eyes		

Table 4

	Similarities Between Gorillas and Humans	Differences Between Gorillas and Humans
Jaw		
Skull		
Spinal column		

Lab 16-1

Primate Characteristics

1. How do the hands of primates aid in their survival?

_____

_____

_____

2. How do the differences between the hands of the gorillas and humans relate to their functions?

_____

_____

_____

3. Is there an advantage to having binocular vision? Explain your answer.

_____

_____

_____

_____

4. What can you infer about gorillas and humans by looking at their jaws, skulls, and spinal columns?

_____

_____

_____

_____

FURTHER EXPLORATIONS

1. Primates possess other unique characteristics including flexible hip and shoulder joints that allow increased arm and leg movement. Research these and other characteristics and find out how they enable primates to better survive in their environments.

2. Find out more about early humanlike primates and how their hands, eyes, skulls, jaws, and teeth were similar or different from those of humans today.

Allelic Frequencies and Sickle-Cell Anemia

Sickle-cell anemia, a potentially fatal disease, results from a mutant allele for hemoglobin, the oxygen-carrying protein in red blood cells. There are two alleles for the production of hemoglobin. Individuals with two Hemoglobin A alleles (AA) have normal red blood cells. Those with two mutant Hemoglobin S alleles (SS) have abnormal sickle-shaped red blood cells and suffer from sickle-cell anemia. Heterozygous (AS) individuals carry the mutant allele but do not suffer from its debilitating effects. They have both normal and sickle-shaped red blood cells.

In the United States, about 1 in 500 African-Americans develops sickle-cell anemia. But in Africa, about 1 in 100 individuals develops the disease. Why is the frequency of a potentially fatal disease so much higher in Africa?

The answer is related to another potentially fatal disease, malaria. Individuals with an AA hemoglobin genotype have a significantly greater risk of contracting malaria and may die from the disease. This results in the removal of Hemoglobin A alleles from the gene pool. The SS genotype, which results in sickle-cell anemia, is usually fatal before the age of twenty. This results in the removal of Hemoglobin S alleles from the population. A person with an AS genotype does not develop sickle-cell anemia and has less chance of contracting malaria. Such a person is better able to survive and reproduce in a malaria-infected region. Therefore, both the individual's A allele and S allele remain in the population. The frequency of the Hemoglobin S allele in malaria-infected regions of Africa is 16%. But in the United States, where malaria has been eradicated, the allelic frequency is 4%.

OBJECTIVES

- Model the effects of natural selection on the frequencies of the Hemoglobin A and S alleles within a population living in a malaria-infected region.
- Explain the process of natural selection as a force affecting allelic frequencies of populations.

MATERIALS

75 red beans and 25 black beans
wax marking pencil
plastic food containers, tall (5)
lid for one of the containers
blindfold
goggles

PROCEDURE

1. Using the wax marking pencil, label one of the containers AA, a second AS, and a third SS, to represent the three possible genotypes for the sickle-cell trait. Label a fourth container "Non-Surviving Alleles." Leave the fifth container unlabeled. It will be used to represent a small population living in a malaria-infected region.

2. Place 75 red beans in the unlabeled container. These beans represent gametes carrying the Hemoglobin A allele.

3. Add 25 black beans to the unlabeled container. These beans represent gametes carrying the Hemoglobin S allele.

4. Close the container with a lid. Shake the container of beans until they are well-mixed.

5. You will now simulate the fusion of gametes and record the resulting genotype of each offspring, using all of the beans in the unlabeled container. You will also simulate the effect of being homozygous or heterozygous for the hemoglobin gene in a malaria-infected region.

Allelic Frequencies and Sickle-Cell Anemia

PROCEDURE continued

Wearing a blindfold to ensure a random choice, one team member will select two beans from the unlabeled container 50 times. After each selection, a second team member will identify and record the genotype of the "offspring" in Table 1 by making a slash under the appropriate column head. The same team member (or another) then places each pair of beans in container AA, AS, or SS, depending on the genotype.

During the periods when the blindfolded team member is making a selection, a third team member will randomly call out the word "malaria" a total of 25 times. (This represents a 50% malaria infection rate.) If the genotype of the selected pair of gametes is AA, that offspring will contract malaria and die. Therefore, place that pair of alleles in the container labeled "Non-Surviving Alleles" and put a circle around the slash (that is, the recorded genotype) in Table 1. If the genotype is AS, the individual will survive. Put a circle around the slash in Table 1, and place the pair of beans in the container labeled AS. If the genotype is SS, the individual will die. As with the individual homozygous for the normal Hemoglobin gene, place the beans in the container labeled "Non-Surviving Alleles" and draw a circle around the recorded genotype.

6. After you have completed the cycle for one generation, you are ready to begin the next. All of the beans in the containers labeled AA and AS should be emptied into the unlabeled container. However, place all the beans from the container labeled SS into the container for "Non-Surviving Alleles." (Since individuals who are homozygous SS will not usually live long enough to have children, you will not use the SS gametes for tallying the next generation.)

7. Count the number of red beans and black beans in the unlabeled container and record the individual and combined totals in the appropriate spaces under Table 1. The combined total represents the total number of alleles for hemoglobin in the population. Calculate the allelic frequencies as shown and record your results.

8. To determine the 50% malaria infection rate for this generation of the population, take the total number of A and S alleles remaining in the population and divide it by 4. This is the number of times you will randomly call out "malaria" in the next round of fusion of gametes. (For example, if you had 47 A and 17 S alleles, you would have a total of 64 alleles in the population. Divide this by 4 and you would get the number 16. This would be 50% of the next generation of people produced by the gamete fusions.)

9. Repeat the procedure you followed in step 5 and then step 6. Use Table 2 to record your data.

10. Repeat the procedure you followed in step 7. Record your data and calculations in the spaces provided under Table 2.

Figure 1

Allelic Frequencies and Sickle-Cell Anemia

DATA AND OBSERVATIONS

Table 1

Second Generation		
AA genotype	**AS genotype**	**SS genotype**

a. How many A alleles are remaining in the population? _____

b. How many S alleles are remaining in the population? _____

c. What is the total number of alleles in the population? A+S = _____

d. What is the frequency (percent) of the A allele? $\frac{A}{A + S} \times 100 =$ _____

e. What is the frequency (percent) of the S allele? $\frac{S}{A + S} \times 100 =$ _____

Table 2

Third Generation		
AA genotype	**AS genotype**	**SS genotype**

a. How many A alleles are remaining in the population? _____

b. How many S alleles are remaining in the population? _____

c. What is the total number of alleles in the population? A + S = _____

d. What is the frequency (percent) of the A allele? $\frac{A}{A + S} \times 100 =$ _____

e. What is the frequency (percent) of the S allele? $\frac{S}{A + S} \times 100 =$ _____

Allelic Frequencies and Sickle-Cell Anemia

Lab 16-2

ANALYSIS

1. a. What was the frequency of the A allele in the original population? _____

   b. What was the frequency of the S allele in the original population? _____

   c. What was the frequency of the A allele in the second generation? _____

   d. What was the frequency of the S allele in the second generation? _____

   e. What was the frequency of the A allele in the third generation? _____

   f. What was the frequency of the S allele in the third generation? _____

   g. Explain your findings. _____

                                        _____

2. Since few people with sickle-cell anemia are likely to survive to have children of their own, why hasn't the Hemoglobin S allele been eliminated by natural selection?

_____

_____

3. Why is the frequency of the Hemoglobin S allele so much lower in the United States than in Africa?

_____

_____

_____

4. Scientists are working on a vaccine against malaria. What impact would the vaccine have on the frequency of the Hemoglobin S allele in Africa?

_____

_____

5. In 5 out of 100 million individuals in a population, the allele for Hemoglobin A will spontaneously undergo mutation into the allele for Hemoglobin S. Will such mutations cause major changes in the allelic frequencies of the population? Explain your answer.

_____

_____

_____

FURTHER EXPLORATIONS

1. Genetic engineering holds the potential for altering genes in human gametes. Write an essay speculating on the impact of genetic engineering on human evolution.

2. Use the library to research a human population that has been geographically or socially isolated over the course of many generations and determine how the population's allelic frequencies have been affected over time.

INVESTIGATION   How Can a Key Be Used to Identify Organisms?

Classification is a way of separating a large group of closely related organisms into smaller subgroups. The scientific names of organisms are based on the classification systems of living organisms. The identification of an organism is easy with a classification system. To identify an organism, scientists often use a key. A key is a listing of characteristics, such as structure and behavior, organized in such a way that an organism can be identified.

OBJECTIVES

- Hypothesize how organisms can be identified with a key.
- Use a key to identify fourteen shark families.
- Examine the method used to make a key.
- Construct your own key that will identify another group of organisms.

MATERIALS

Figure 2
Key on page 103

PROCEDURE

1. Make a **hypothesis** to describe how sharks can be identified using a key. Write your hypothesis in the space provided.

2. Use Figure 1 as a guide to the shark parts used in the key on page 119.

3. Read statements 1A and 1B of the key. They describe a shark characteristic that can be used to separate the sharks into two major groups. Then study Shark 1 in Figure 2 for the characteristic referred to in 1A and 1B. Follow the directions in these statements and continue until a family

name for Shark 1 is determined. For example, to key a shark that has a body that is not kite shaped and has a pelvic fin and six gill slits, follow the directions of 1B and go directly to statements 2. Follow statement 2B to statements 3. At statement 3A, identify the shark as belonging to Family Hexanchidae.

4. Continue keying each shark until all have been identified. Write the family name on the line below each shark in Figure 2.

5. Have the teacher check your answers.

HYPOTHESIS

_____

_____

Figure 1

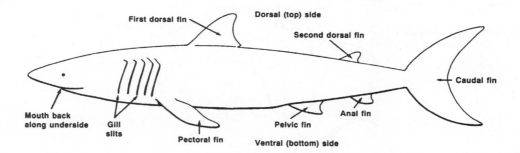

How Can a Key Be Used to Identify Organisms?

Figure 2

1. _____

2. _____

3. _____

4. _____

5. _____

6. _____

7. _____

8. _____

9. _____

10. _____

11. _____

12. _____

13. _____

14. _____

How Can a Key Be Used to Identify Organisms?

1. A. Body kitelike in shape (if viewed from above) .... Go to statement 12
   B. Body not kitelike in shape (if viewed from above) .... Go to statement 2

2. A. Pelvic fin absent and nose sawlike .... Family Pristophoridae
   B. Pelvic fin present .... Go to statement 3

3. A. Six gill slits present .... Family Hexanchidae
   B. Five gill slits present .... Go to statement 4

4. A. Only one dorsal fin present .... Family Scyliorhinidae
   B. Two dorsal fins present .... Go to statement 5

5. A. Mouth at front of head rather than back along underside of head ...... Family Rhinocodontidae
   B. Mouth back along underside of head .... Go to statement 6

6. A. Head expanded on side with eyes at end of expansion .... Family Sphyrnidae
   B. Head not expanded .... Go to statement 7

7. A. Top half of caudal fin exactly same size and shape as bottom half .... Family Isuridae
   B. Top half of caudal fin different in size and shape from bottom half .... Go to statement 8

8. A. First dorsal fin very long, almost half total length of body .... Family Pseudotriakidae
   B. First dorsal fin length much less than half total length of body .... Go to statement 9

9. A. Caudal fin very long, almost as long as entire body .... Family Alopiidae
   B. Caudal fin length much less than length of entire body .... Go to statement 10

10. A. Nose with long needlelike point on end .... Family Scapanorhynchidae
    B. Nose without needlelike point .... Go to statement 11

11. A. Anal fin absent .... Family Squalidae
    B. Anal fin present .... Family Carcharhinidae

12. A. Small dorsal fin present near tip of tail .... Family Rajidae
    B. Small dorsal fin absent near tip of tail .... Go to statement 13

13. A. Hornlike appendages at front of shark .... Family Mobulidae
    B. Hornlike appendages not present at front of shark .... Family Dasyatidae

ANALYSIS

1. What is a classification key and how is it used?

2. List four different characteristics that were used in the shark key.

3. a. Which main characteristic could be used to distinguish shark 4 from shark 8?

   b. Which main characteristic could be used to distinguish shark 4 from shark 7?

How Can a Key Be Used to Identify Organisms?

ANALYSIS continued

4. Prepare your own key for the five fish in Figure 3. Use the same format as on page 119. The family names to be used are the numbers I, II, III, IV, and V. Your key should correctly use traits that will lead to each fish family. To help you get started, the first statements are given. Statement 1 divides the five fish into two main groups, based on body shape. Next, choose another characteristic that will divide the fish not having a tubelike body into two groups. Continue to choose characteristics that will separate a group into smaller groups. Write your key in the space below.

Figure 3

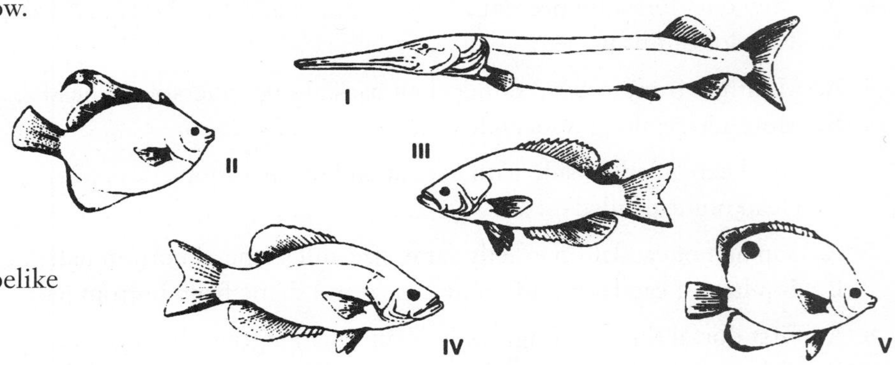

1. A. Fish with long tubelike body
   B. Fish with body shape not tubelike

Key

1. A. _____
   B. _____

2. A. _____
   B. _____

3. A. _____
   B. _____

4. A. _____
   B. _____

CHECKING YOUR HYPOTHESIS

Did your **hypothesis** describe the key correctly?

_____

_____

FURTHER INVESTIGATIONS

1. Exchange keys with a classmate. Work through it to identify the fish. Is the key correct?

2. The library will have many books that include simple keys to different plants and animals, as well as to rocks, fossils, and stars. Select a book that includes keys to local plants or animals. Take a walk and practice using the key to identify some of the organisms that live in your area.

Comparing Characteristics of Organisms

Lab

17-2

Living things are all around us. In order to identify organisms effectively, scientists have developed methods of classifying organisms into six groups called kingdoms. The kingdoms reflect evolutionary relationships among the organisms and, in general, are distinguished by differences in the organisms' cellular characteristics and methods of obtaining energy. For example, some organisms are prokaryotes, which means that their cells lack a distinct nucleus as well as other membrane-bound organelles. Other organisms are eukaryotes, which do have a nucleus and other organelles. Some types of organisms, called autotrophs, make their own food. Others, called heterotrophs, must take in food. In addition, organisms can be unicellular or multicellular. Also, organisms may or may not be able to move from place to place.

OBJECTIVES

- Observe and compare the characteristics of organisms from the six kingdoms of life.
- Identify some distinguishing features of the six kingdoms.

- Classify organisms into one of the six kingdoms.

MATERIALS

stations with:
    examples of archaebacteria
    examples of eubacteria
    examples of protists
    examples of fungi

    examples of plants
    examples of animals
your textbook
reference books
goggles

PROCEDURE

1. Your teacher has set up six exploration stations around the classroom. At each station, you will find a number of organisms from one of the kingdoms, which is not identified.

2. As a group, move to your assigned station and observe the organisms present. Your teacher may have set up slides mounted on microscopes for you to observe either living or preserved specimens. Use proper microscope techniques to examine these organisms.

3. Some organisms are macroscopic; they can be viewed without using a microscope. **CAUTION:** *Handle all living organisms carefully.* Wash your hands after handling specimens.

4. While at the station, record in one of the tables in Data and Observations all of your observations of the organisms. Record the station number at the top of the table. Based on your observations, identify the kingdom to which the organisms at the station belong. Record the name of the kingdom at the top of the table. You may use your textbook and other reference books to complete the table if needed.

5. When told to do so, move to the next station and observe organisms of another kingdom. Repeat step 4.

Comparing Characteristics of Organisms

DATA AND OBSERVATIONS

Table 1

		Station # _____ : Kingdom _____			
Example	Micro-scopic?	Prokaryote or Eukaryote	Multicellular or Unicellular	Autotrophic or Heterotrophic	Distinguishing Features

Table 2

		Station # _____ : Kingdom _____			
Example	Micro-scopic?	Prokaryote or Eukaryote	Multicellular or Unicellular	Autotrophic or Heterotrophic	Distinguishing Features

Table 3

		Station # _____ : Kingdom _____			
Example	Micro-scopic?	Prokaryote or Eukaryote	Multicellular or Unicellular	Autotrophic or Heterotrophic	Distinguishing Features

Comparing Characteristics of Organisms

Table 4

Station # _____ : Kingdom _____					
Example	Micro-scopic?	Prokaryote or Eukaryote	Multicellular or Unicellular	Autotrophic or Heterotrophic	Distinguishing Features

Table 5

Station # _____ : Kingdom _____					
Example	Micro-scopic?	Prokaryote or Eukaryote	Multicellular or Unicellular	Autotrophic or Heterotrophic	Distinguishing Features

Table 6

Station # _____ : Kingdom _____					
Example	Micro-scopic?	Prokaryote or Eukaryote	Multicellular or Unicellular	Autotrophic or Heterotrophic	Distinguishing Features

Comparing Characteristics of Organisms

ANALYSIS

1. What features do organisms in each of the kingdoms have in common? List the features in Table 7 below.

Table 7

Kingdom	Common Features of the Organisms
Archaebacteria	
Eubacteria	
Protista	
Fungi	
Plantae	
Animalia	

2. Which kingdom(s) are made up of prokaryotes? How do you know?

_____

_____

3. Which kingdom(s) include both autotrophs and heterotrophs?

_____

4. Which kingdom(s) include both unicellular and multicellular organisms?

_____

5. Which kingdom(s) are made up of multicellular heterotrophs only?

_____

6. Scientists have used various methods to classify living things. Which method of classification did you use in this Exploration? Explain your answer.

_____

_____

FURTHER EXPLORATIONS

1. How would you classify an unknown organism into one of the six kingdoms of life? Using your textbook and other reference materials, write a series of questions that would enable you to systematically classify an unknown organism.

2. Continue your exploration of classification by further classifying some of the organisms you observed in each of the stations into smaller groups, or taxonomic levels.

Lab 18-1

EXPLORATION

Viral Replication

Viruses are very successful at invading the cells of organisms. A virus first becomes attached securely to the outside of a host cell. Then, some viruses inject their nucleic acid into the cell and leave their coat outside the cell; other viruses still have their coats when they enter the cell. Once many types of viruses are inside a host cell, they insert viral genes into the host DNA. The viral genes direct replication, causing the cell to make many new copies of viral genes. At the same time, the cell's protein synthesis machinery is directed to make many new viral coats and enzymes. The virus particles are assembled, and the new viruses escape from the cell either by exocytosis or by bursting out of the cell.

OBJECTIVES

- Trace the steps of viral replication in cells.
- Describe the steps of viral replication.
- Construct an analogy for viral replication in cells.

MATERIALS

glue
scissors
photocopy of model viruses and viral parts
goggles

PROCEDURE

1. Study Figures 1 and 2 so that you become familiar with viral structures and with the models of viral structures that you will use in this Exploration.

2. The teacher will give you a photocopy of drawings that represent viruses and viral parts. Cut out the drawings.

3. Each drawing corresponds to one of the numbered steps of viral replication listed at the end of the Procedure. Using Figure 2 as a key, label the tab on each drawing with the number of the step of viral replication that it represents.

4. Examine Figure 3 under Data and Observations. The diagram summarizes the steps of viral replication within a cell. The location of each step is labeled. Match each cutout drawing with the correct label. Then copy the numbers from the drawings onto the appropriate spaces provided in the diagram.

5. Place each drawing over its corresponding label in Figure 3. When complete, have your diagram checked by the teacher for accuracy.

6. Once the drawings are in their proper positions on the diagram, glue them in place.

Copyright © Glencoe/McGraw-Hill, a division of The McGraw-Hill Companies, Inc.

Viral Replication

PROCEDURE continued

Figure 1

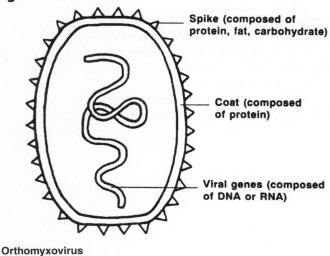

Spike (composed of
protein, fat, carbohydrate)

Coat (composed
of protein)

Viral genes (composed
of DNA or RNA)

Orthomyxovirus
× 20 000

Viral structure

Figure 2

Model
viral coat

Model
viral genes

Model
viral genes in coat

Key for cutout drawings

Steps of Viral Replication

1. The virus attaches to the host's cell membrane.

2. The virus enters the cell.

3. The protein coat of the virus is removed.

4. Viral genes are activated.

5. a. The activated genes direct replication of viral genes.

   b. The activated genes direct protein synthesis to make
   new viral coats and enzymes.

6. a. New viral genes are completed.

   b. New viral coats are completed.

7. The new viruses are assembled. Genes are inserted
   into the protein coats.

8. The new viruses are released from the cell.

Viral Replication

DATA AND OBSERVATIONS

Life Cycle of a Virus

Figure 3

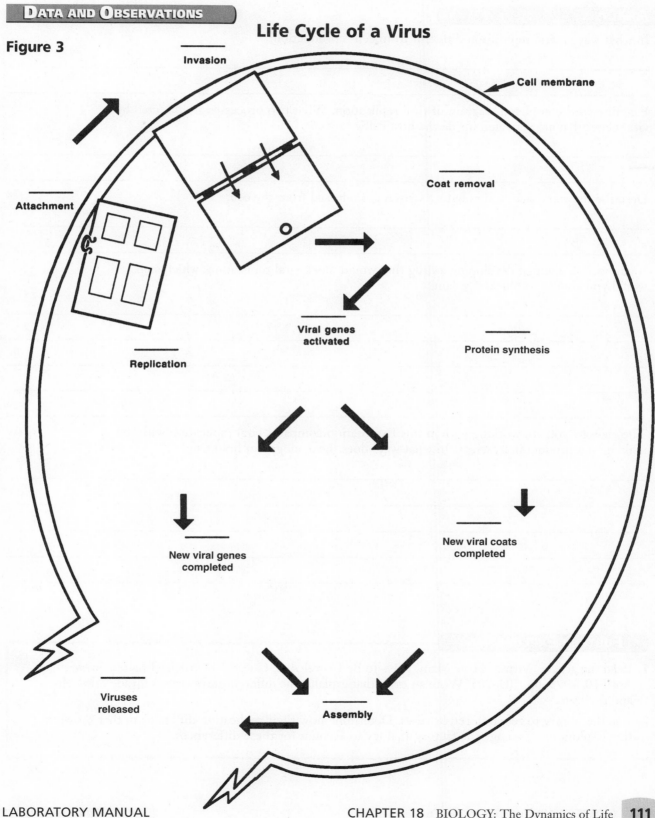

Invasion

Cell membrane

Coat removal

Attachment

Viral genes
activated

Protein synthesis

Replication

New viral genes
completed

New viral coats
completed

Viruses
released

Assembly

Viral Replication

ANALYSIS

1. In what way is viral replication different from cell reproduction?

_____

_____

2. Examine your completed diagram of viral replication. What two processes are directed by viral genes that are activated inside the host cell?

_____

_____

3. Describe the stage that occurs before viruses are released from the cell.

_____

_____

4. If you were a scientist developing a drug that would block viral replication, which steps would you choose to block? Explain.

_____

_____

_____

_____

_____

5. The analogy you are working with in this Exploration compares viral replication with the making of a product in a factory. In what ways does the analogy not hold true?

_____

_____

_____

_____

_____

FURTHER EXPLORATIONS

1. Read the article "Viruses Have Many Ways to Be Unwelcome Guests," by Michael Balter, *Science*, April 10, 1998, pp. 204–205. Write an essay that explains the different ways viruses invade host cells and replicate.

2. Use the library to research retroviruses. Determine how their replication differs from that shown in this Exploration. Change the factory analogy to account for these differences.

Name   Date   Class

INVESTIGATION

How Are Bacteria Affected by Heat?

Bacteria comprise much of the dry weight of feces and can contaminate lakes, streams, and groundwater through untreated sewage or drainage from cattle farms. Because of this potential for contamination, water is usually treated with chlorine. However, milk, another potential source of bacteria, cannot be treated with chlorine. Raw milk—which is milk taken directly from the cow, then filtered and consumed—was once a cause of illness. Then in the late 1800s, a technique for destroying various bacteria in milk by heating was developed by the chemist Louis Pasteur. Modern-day pasteurization consists of heating milk to 145 degrees Fahrenheit (about 63° C) for 30 minutes. The heat kills many of the bacteria present in the milk. The milk is then cooled and kept at a temperature below 50 degrees Fahrenheit to prevent remaining bacteria from spoiling the milk. There is a "shelf life" for milk, after which it can spoil due to a variety of bacteria. In this lab, you will investigate how heat affects *Escherichia coli* (or *E. coli*), one of the types of bacteria found in milk.

OBJECTIVES

- Practice sterile techniques for handling bacterial cultures and inoculating agar plates.
- Hypothesize how heat affects the growth of bacteria.
- Observe the effects of heat on the growth of bacteria.

MATERIALS

disinfectant solution
sterile petri dishes containing hardened bacto-methylene blue agar (2)
sterile cotton swabs (10)
test tube containing 10 mL of *E. coli* culture
test tube with 10 mL of milk inoculated with *E. coli* and autoclaved
hot plate
water bath
thermometer
test tubes (3)
test-tube holder
test-tube rack

sterile pipette
wax marking pencil
incubator
refrigerated milk
paper towels (2)
tape, masking
250-mL beaker with 100 mL of alcohol
25-mL graduated cylinder
clock or watch
thermal mitts
laboratory apron
goggles

PROCEDURE

CAUTION: *Do not touch eyes, mouth, or any other part of your face while doing this lab. Wear your laboratory apron and goggles. Wash your work surface with disinfectant solution, using a paper towel, both before and after doing the lab.*

How Are Bacteria Affected by Heat?

PROCEDURE continued

Part A. Preparing the Treatments

1. Prepare a hot water bath by heating the water to 63°C. One person should monitor the hot plate and thermometer while the rest of the team does the following activities.

2. Place the two petri dishes containing bacto-methylene blue agar on the counter in front of you. Turn the dishes upside down, being careful not to allow them to open. With a wax marking pencil, divide each dish into thirds as shown in Figure 1. On dish A, number the sections 1, 2, and 3. On dish B, number the sections 4, 5, and 6. These numbers refer to the following treatments:
   1. Control
   2. Refrigerated milk
   3. Refrigerated milk inoculated with *E. coli*
   4. Refrigerated milk inoculated with *E. coli* and autoclaved
   5. *E. coli* culture
   6. Refrigerated milk inoculated with *E. coli* and heated on the hot plate

3. Label 3 test tubes with numbers 2, 3, and 6 to correspond to the treatments listed in step 2, and place them in a test-tube rack. To each test tube, add 10 mL of refrigerated milk from the same milk source.

4. Label with the number 4 the test tube of milk that has been inoculated and autoclaved. Label with the number 5 the test tube containing 10 mL of *E. coli* culture.

5. With a sterile pipette, add 1 mL of *E. coli* culture from test tube 5 to test tubes 3 and 6. Place the contaminated pipette in the container of alcohol on the counter. Carefully roll the test tubes

between your palms to mix the *E. coli* into the milk. Do not tip the test tubes. Put the test tubes back into the rack after each mixing.

6. Using a test-tube holder, place test tube 6 in the hot water bath. Insert the thermometer into the test tube, and heat the tube for 30 minutes or until the milk is 63°C. Using the test tube holder, remove test tube 6 from the heat and place it into the test-tube rack. Place the contaminated thermometer in the container of alcohol.

Part B. Inoculating the Agar Plates

1. Dip a sterile cotton swab into test tube 2. Press the swab against the inside of the tube to remove excess milk. Lift the lid off petri dish A, and touch the agar surface in section 2 with the cotton swab. Discard the swab in the beaker of alcohol. Using another sterile swab, lightly spread the milk in section 2, using a tight s-shaped pattern. Replace the lid. Discard the swab in the alcohol.

2. Dip a sterile cotton swab into test tube 3. Press the swab against the inside of the test tube to remove excess milk. Lift the lid off petri dish A and touch the agar surface in section 3 with the cotton swab. Discard the swab in the beaker of alcohol. **CAUTION:** *Do not let the cotton end of the swab touch the outside of the dish or any other surface. If any of the liquid spills from test tube 3, call your teacher immediately.* Using another sterile swab, lightly spread the culture in section 3, using a tight s-shaped pattern. Replace the lid on the petri dish and discard the cotton swab. Secure the lid of the petri dish with tape, label the tape with the name of your team, and turn the dish upside down.

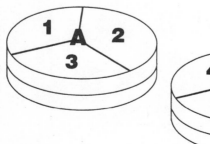
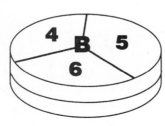

Figure 1

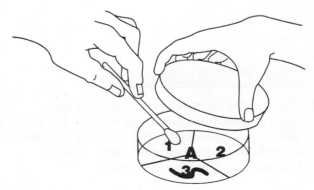

How Are Bacteria Affected by Heat?

3. You will now inoculate petri dish B. Dip a cotton swab into test tube 4. Press the swab against the inside of the tube to remove excess milk. Lift the lid, and touch the agar surface in section 4 with the cotton swab. Discard the swab in the alcohol. Using another sterile swab, spread the treatment throughout section 4. Replace the lid on the petri dish and discard the swab.

4. Repeat step 3 with test tubes 5 and 6, using sterile cotton swabs to apply and spread cultures in sections 5 and 6 of petri dish B. Exercise special caution when inoculating from test tube 5. When you have completed the inoculations, secure the lid of the petri dish with tape. Label the tape with the name of your team, and turn the dish upside down.

5. Place both petri dishes in the incubator overnight at 35–37°C. Incubate the dishes upside down to prevent moisture from accumulating on the surface of the agar. Wash your hands thoroughly.

6. Make a **hypothesis** to describe how heat will affect the bacteria. Write your hypothesis in the space provided.

7. On the following day or during the next lab period, examine the two petri dishes and record your observations in Table 1. *E. coli* colonies will appear dark and buttonlike, often with concentric rings and a greenish metallic sheen. Other bacteria may also be present but will not have the same appearance.

8. After completing your observations, dispose of the petri dishes as instructed by the teacher.

HYPOTHESIS

_____

_____

DATA AND OBSERVATIONS

Table 1

Treatment	Observations
#1 Control	
#2 Refrigerated milk	
#3 Refrigerated milk + *E. coli*	
#4 Refrigerated milk + *E. coli*, autoclaved	
#5 *E. coli*	
#6 Refrigerated milk + *E. coli*, heated	

How Are Bacteria Affected by Heat?

ANALYSIS

1. What was the purpose of the control section on the agar?

_____

_____

2. What was the purpose of section 2 on the agar?

_____

_____

3. What was the purpose of section 5?

_____

_____

4. Compare and explain the results in sections 3 and 5.

_____

_____

5. Explain the results in section 6.

_____

_____

6. Explain the results in section 4.

_____

_____

_____

CHECKING YOUR HYPOTHESIS

Was your **hypothesis** supported by your data? Why or why not?

_____

_____

FURTHER INVESTIGATIONS

1. Design an experiment that tests the effects of other factors on bacterial growth. Have the teacher approve your experimental design before you conduct it.

2. Go to the library and research the use of *E. coli* in recombinant DNA experiments. When *E. coli* was first used for this type of research, many people feared its use. Explain why people thought that and why they need not have been worried.

Lab
19-1

INVESTIGATION **How Can Digestion Be Observed in Protozoans?**

Chemical digestion of food is a complex series of chemical reactions, each of which is controlled by a different enzyme. Each enzyme works best under a different set of conditions. As the pH of the food changes during the digestion process, enzymes with different optimal pH ranges become active. As food is broken down in the food vacuoles of *Paramecium* or other ciliates, the pH of the vacuole contents changes. If the vacuole contents are stained with a pH indicator, the progress of digestion can be followed.

OBJECTIVES

- Prepare an indicator-stained food source for *Paramecium*.
- Observe the feeding behavior of *Paramecium*.

- Hypothesize how the indicator-stained food will change color as is digested by the *Paramecium*.

MATERIALS

small beaker
cream or whole milk
wood splint
Congo red indicator
microscope slide
coverslip
toothpick
petroleum jelly

droppers (3)
methyl cellulose
culture of *Paramecium*
compound light microscope
clock or stopwatch
laboratory apron
goggles

PROCEDURE

Part A. Preparing Materials

1. Obtain a small amount of milk in a beaker.

2. Use a wood splint to transfer a few grains of Congo red to the milk. Use caution when working with Congo red. Stir the mixture with the wood splint. Congo red is used to stain structures that contain fats or oils. Congo red is also an indicator of pH. The indicator is red at a pH of 5.0 or greater. It is blue at a pH of 3.0 or less.

3. Use a toothpick to outline a square of petroleum jelly on the center of a microscope slide as shown in Figure 1. The square should be the size of a coverslip.

4. Use a dropper to place one drop of methyl cellulose in the center of the square. This substance will slow the movement of the paramecia.

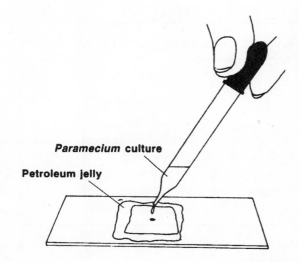

Paramecium culture
Petroleum jelly

Figure 1

How Can Digestion Be Observed in Protozoans?

PROCEDURE continued

5. Use a clean dropper to add one drop of *Paramecium* culture to the slide.

6. Use a clean dropper to add one drop of stained milk to the slide.

7. Add a coverslip to the slide.

8. Make a **hypothesis** about how the color of the indicator-stained food will change as it is digested by the paramecium. Write your hypothesis in the space provided.

Part B. Observing Digestion

1. Observe your slide under low power. Locate a paramecium that is near some of the milk's stained fat droplets.

2. Change to high power, and look for food vacuoles in the paramecium. Record the color of the contents of the food vacuoles. The color is an indication of the contents' pH. Write your

observations of the paramecium's behavior and the movement of its food vacuoles in Table 1.

3. In Table 2, make a drawing of the paramecium. Label the oral groove, gullet, and any food vacuoles that you see. Use your textbook for more information on the structures of *Paramecium*.

4. Continue to observe the paramecium. Every ten minutes for one-half hour, record the color of the food vacuole contents and your observations, and make a drawing of the paramecium. Be careful to draw the positions of the food vacuoles accurately.

HYPOTHESIS

_____

_____

DATA AND OBSERVATIONS

Table 1

Time (min)	Color of food vacuole contents	Other observations
0		
10		
20		
30		

How Can Digestion Be Observed in Protozoans?

Table 2

Drawings of *Paramecium*			
0 min	10 min	20 min	30 min

ANALYSIS

1. What color was the mixture of Congo red and milk? _____ What does this tell you about the pH of the milk?

_____

_____

2. When you first observed the paramecium, did it contain any stained food vacuoles? _____ If not, when did you first see stained food vacuoles?

_____

_____

3. What color were the contents of the food vacuoles when the paramecium first began to digest the milk?

_____

What does this indicate about the pH of the food at the beginning of digestion?

_____

_____

How Can Digestion Be Observed in Protozoans?

ANALYSIS continued

4. At what time did the contents of the food vacuoles change color? _____ What color did

they become? _____ What does this indicate about the pH of the food vacuole

contents as digestion progresses?

_____

_____

_____

5. Did the contents of the food vacuoles change color a second time? _____ What does the

change or lack of change indicate about the pH of the food vacuole contents late in digestion?

_____

_____

_____

6. Thymol blue is an indicator that is red at a pH of 1.2 or less. It is yellow between 2.8 and 8.2. It is blue at
10.0 or higher. How might repeating the experiment with this indicator give additional information about
the pH of food as it is digested in the food vacuoles?

_____

_____

_____

CHECKING YOUR HYPOTHESIS

Was your **hypothesis** supported by your data? Why or why not?

_____

_____

_____

FURTHER INVESTIGATIONS

1. Repeat the experiment with another kind of ciliate, such as *Blepharisma* or *Bursaria*.

2. Repeat the experiment with the addition of a drop of *Didinium*. This ciliate is a predator that feeds on
Paramecium. See if the *Didinium* shows color changes as it digests the stained *Paramecium*.

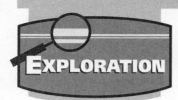

EXPLORATION · Observing Algae

Algae are simple plantlike organisms belonging to the Kingdom Protista. Algae contain chlorophyll and are therefore capable of carrying out photosynthesis. The chlorophyll is contained in organelles called chloroplasts. Specialized structures within the chloroplasts of some algae are important for starch formation and storage. These structures are known as pyrenoids. The cellular organization of algae varies. Some are unicellular, while others are multicellular. The multicellular algae may be in the form of seaweeds, threads called filaments, or groups called colonies. Colonies vary in size from a few cells to thousands of cells, depending on the species.

OBJECTIVES

- Prepare wet mounts of algae for microscopic examination.
- Locate and identify chloroplasts, holdfasts, and pyrenoids of various species of algae.

- Distinguish among unicellular, filamentous, and colonial forms of organization in algae.

MATERIALS

compound light microscope
microscope slides (6)
coverslips (6)
droppers (6)
iodine solution
paper towel
laboratory apron

goggles
Living specimens of
    Closterium
    Oedogonium
    Scenedesmus
    Synedra
    Ulothrix
    Volvox

PROCEDURE

Part A. Unicellular Algae

1. Remove a small sample of *Synedra* from the specimen jar with a dropper. Place a drop of the culture on a clean microscope slide and cover with a coverslip.

2. Locate the alga under low power of the microscope. Look for an elongated cell as shown in Figure 1.

3. Switch to high power and note the cell wall. Label the cell wall of the cell in Figure 1.

4. Record in Table 1 the cellular organization of this alga and its number of cells.

5. Prepare a wet mount, as in step 1, of *Closterium*, using a clean dropper, slide, and coverslip.

6. Observe first under low power and look for a yellow-green crescent-shaped cell, as shown in Figure 2. The cell has a constriction in its middle called an isthmus.

Synedra

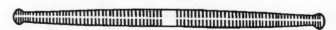

Figure 1

Observing Algae

PROCEDURE continued

7. Add a drop of iodine to the edge of the coverslip. Using a piece of paper towel as in Figure 3, carefully draw the iodine across the slide. The iodine will stain pyrenoids blue-black. **CAUTION:** *If iodine spillage occurs, wash with water and call the teacher immediately.*

8. Switch to high power and examine the alga again. Label the cell wall, pyrenoid, and isthmus in Figure 2.

9. Record in Table 1 the cellular organization of this alga and its number of cells..

Part B. Filamentous Algae

Oedogonium is a pale yellow-green alga. It may be floating or attached to surfaces in the specimen jar by means of a holdfast cell located at its base.

1. Locate the alga in the jar and remove a small sample with a clean dropper.

2. Place the specimen on a clean slide and cover with a coverslip.

3. Observe the alga under low power. Look for an unbranched filament as shown in Figure 4.

4. Stain the alga with iodine, using the technique in step 7 of Part A.

5. Observe the stained alga again under low power. Note the netlike chloroplast and the many pyrenoids.

6. Switch to high power and observe the cell wall, chloroplast, pyrenoid, and holdfast cell. Label these structures in Figure 4.

7. Record the cellular organization of this alga in Table 1. Estimate and record the number of individual cells that compose the filament.

Ulothrix is an alga usually found attached by its holdfast cell to the sides of a culture jar or to substances in the jar.

8. Locate the alga and remove a small sample from the jar with a clean dropper. Make a wet mount of this alga with a clean slide and coverslip.

9. Look for an unbranched filament as shown in Figure 5. Observe it first under low power. Note the ribbon-shaped chloroplast.

Figure 2

Closterium

Figure 3

Figure 4

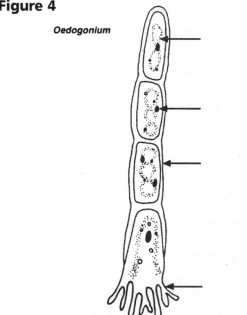

Oedogonium

10. Stain with iodine, using the procedure described in step 7 of Part A.

11. Switch to high power. Find the cell wall, chloroplast, pyrenoid, and holdfast. Label these structures in Figure 5.

12. Record this alga's cellular organization in Table 1. Estimate and record the number of individual cells that compose the filament.

Observing Algae

Figure 5

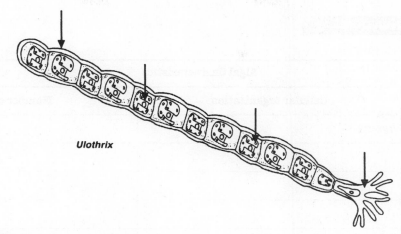

Ulothrix

Part C. Colonial Algae

Volvox can be seen as tiny green spheres spinning through the water. A *Volvox* colony consists of a single layer of 500 to 60 000 cells arranged in a hollow sphere.

1. Locate and remove a colony of *Volvox*, using a clean dropper. Make a wet mount of this alga, using a clean slide and coverslip.

2. Look for a large, green spherical colony as shown in Figure 6. Observe first under low power.

3. Switch to high power. Locate a daughter colony inside the sphere. Daughter colonies are groups of cells within the colony that eventually break away to form new colonies. Label the flagella and daughter colonies in Figure 6.

4. Record this alga's cellular organization in Table 1. Estimate and record the number of individual cells that make up the colony.

Scenedesmus can be found floating throughout the water in the specimen jar.

5. Prepare a wet mount of *Scenedesmus*. Be sure to use a clean dropper, slide, and coverslip.

6. Look for a colony of oval or crescent-shaped cells with short spines as shown in Figure 7. Observe first under low power.

7. Switch to high power and find the cell wall, chloroplasts, and spines. Label these structures in Figure 7.

8. Record this alga's cellular organization in Table 1. Estimate and record the number of cells that make up the colony.

Figure 6

Volvox

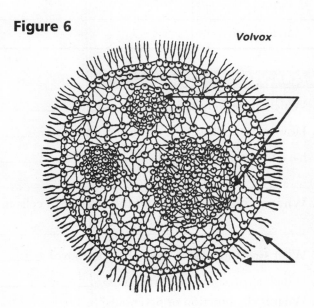

Figure 7

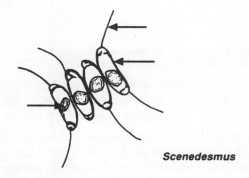

Scenedesmus

Copyright © Glencoe/McGraw-Hill, a division of The McGraw-Hill Companies, Inc.

Observing Algae

DATA AND OBSERVATIONS

Table 1

Algal Characteristics		
Algal specimen	**Cellular organization**	**Number of cells**
Synedra		
Closterium		
Oedogonium		
Ulothrix		
Volvox		
Scenedesmus		

ANALYSIS

1. Which algae were unicellular? _____

2. How many cells were present in *Scenedesmus*? _____ Did all of the specimens

have the same number of cells? _____ Explain.

_____

3. Why was it difficult to count the number of cells in *Volvox*?

_____

What structures allow this colony to move? _____

4. Which algae contained pyrenoids? _____

5. What is the function of pyrenoids?

_____

6. Why was iodine used to locate pyrenoids?

_____

FURTHER EXPLORATIONS

1. Obtain a sample of pond water. Prepare a slide for microscopic examination and locate algae. Look carefully since many different types of algae will be present. Make a drawing of each type of alga observed and identify as many as possible.

2. Carageenan and alginate are polysaccharides derived from the cell walls of brown algae and red algae. They are used as thickeners in ice cream and some cheeses, as well as in cosmetics. Visit a local grocery store and look for these substances on the labels of various products. Make a list of the products and the algae-derived substances they contain.

EXPLORATION Identification of Common Fungi

Fungi are classified mainly by means of their reproductive structures. Zygomycotes produce asexual spores in a sporangium. During sexual reproduction, zygomygotes produces thick-walled zygospores. Ascomycotes produce sexual spores in a sac called an ascus. They also produce asexual spores called conidia. Basidiomycotes produce sexual spores in a club-shaped structure called a basidium. Basidia are arranged on the gills in the cap of the fungus. Some basidiomycotes also reproduce asexually by producing conidia. Deuteromycotes have only an asexual reproductive phase.

OBJECTIVES

- Observe the color and appearance of fungus colonies.
- Draw and label the reproductive structures of various types of fungi.
- Identify the division to which the different types of fungi belong.

PROCEDURE

1. Examine a fungus culture and record in Table 1 the color and appearance of the colony.

2. Gently touch the reproductive surface of the fungus colony with the adhesive side of a piece of cellophane tape. Do not crush the reproductive structures. The teacher will demonstrate where on the fungus the spores are located.

3. Carefully place the tape, adhesive side up, on a microscope slide.

4. Observe the reproductive structures under low power of the microscope.

5. Draw the structures in the space provided in Table 1. Use the drawings in Figure 1 to label the structures.

6. Identify the division to which the fungus belongs by using your textbook or any other available reference materials. Record the division in Table 1.

7. Repeat this procedure for each of the fungi provided.

8. Wash your hands after handling fungi.

MATERIALS

cultures of fungi (4)   microscope slides (4)
cellophane tape   laboratory apron
compound light   goggles
  microscope

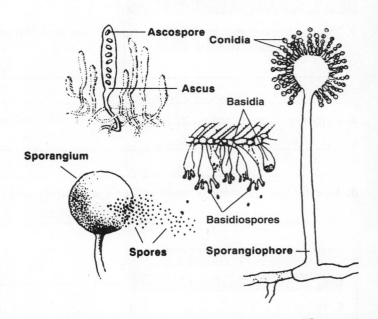

Figure 1

Identification of Common Fungi

DATA AND OBSERVATIONS

Table 1

Identification of Common Fungi			
Fungus	Color and appearance of colony	Drawings of reproductive structures	Division

ANALYSIS

1. Write a general description of the spores you observed in this Exploration. Include relative numbers, shape, and size.

_____

_____

2. Why are fungi identified by using entire reproductive structures rather than by using only spores?

_____

_____

3. List the four divisions of fungi and give an example of a fungus from each division.

_____

_____

FURTHER EXPLORATIONS

1. Bring some fungus specimens from home, and follow the procedure you used in the Exploration to identify the division to which they belong. Good places to look for fungi are on cheese and fruit and in the spaces between tiles in the shower.

2. Using spores from one of the fungus cultures in this Exploration, inoculate a petri dish containing nutrient medium. Study the growth of the fungus and describe its life cycle.

What Conditions Does Bread Mold Need for Growth?

Molds are organisms that belong to Kingdom Fungi. Molds therefore have characteristics different from bacteria and protists. First, fungi are usually multicellular and often can be seen without a microscope. Fungus cells have nuclei. Fungi contain no chlorophyll and so are not able to make their own food. Organisms in this kingdom are either saprophytes, mutualists, or parasites. Saprophytic fungi obtain their food by feeding on once-living material. Mutualistic fungi live in close association with other organisms, such as algae. Parasitic fungi obtain their food by feeding on living material.

OBJECTIVES

- Examine bread mold by using a hand lens or stereomicroscope.
- Hypothesize the conditions a bread mold needs for growth.
- Determine if bread mold can use a variety of food sources.
- Determine if bread mold needs moisture.

MATERIALS

hand lens or stereomicroscope
petri dish containing living bread mold
small jars with covers (6)
water
wax marking pencil
cardboard from a box (2 pieces)
dehydrated potato flakes
raisins
cotton swabs (6)
laboratory apron
goggles

PROCEDURE

Part A. Observing Bread Mold

1. Use a hand lens or stereomicroscope to observe bread mold growing in a petri dish.

2. Note the following parts.
   (a) Tiny black, round structures called sporangia are located on top of long stalks called sporangiophores. Reproductive spores form in these structures.
   (b) A mass of threadlike structures spreads along the surface of the mold's food supply. These structures are called hyphae, and they enable the mold to secure its food.
   (c) Hyphae that penetrate the food supply are rhizoids.

3. Label the following parts on Figure 1:
   sporangium, sporangiophore, rhizoid, hypha.

Part B. Testing Different Possible Sources of Food

1. Label six small jars with your name and the numbers 1 to 6.

2. Prepare the jars as follows:
   Jar 1: Cover the bottom with dehydrated (dry) potato flakes.
   Jar 2: Cover the bottom with potato flakes and add enough water to make a paste.
   Jar 3: Cover the bottom with dry raisins.
   Jar 4: Cover the bottom with dry raisins and add enough water to soak them thoroughly.
   Jar 5: Stand a small piece of cardboard upright in the jar.
   Jar 6: Stand a small piece of cardboard upright in the jar and add a small amount of water to the bottom of the jar.

3. Rub a damp cotton swab over the surface of the bread mold studied in Part A. Rub the swab over the surface of the contents of jar 1. Repeat this procedure for the remaining five jars, using a different swab for each jar.

4. Cover the jars and let them sit for several days at room temperature.

What Conditions Does Bread Mold Need for Growth?

PROCEDURE continued

5. Make a **hypothesis** to describe the conditions a mold needs to grow. Write your hypothesis in the space provided.

6. After several days, examine each jar for the presence or absence of bread-mold growth.

7. Record your observations in Table 1.

8. Dispose of your jars according to the teacher's instructions.

9. Wash your hands when you finish.

HYPOTHESIS

_____

_____

DATA AND OBSERVATIONS

Table 1

Jar	Contents	Mold growth?
1	Dry potato flakes	
2	Wet potato flakes	
3	Dry raisins	
4	Wet raisins	
5	Dry cardboard	
6	Wet cardboard	

Results of Tests

Figure 1

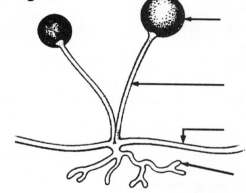

ANALYSIS

1. In which jars did bread mold grow? _____

2. What growth conditions were supplied to the fungus in the jars in which you observed bread mold growth?

_____

3. How does moisture affect the growth of bread mold? _____

CHECKING YOUR HYPOTHESIS

Was your **hypothesis** supported by your data? Why or why not?

_____

FURTHER INVESTIGATIONS

1. Set up two additional jars with the same contents as the jar that had the best growth of fungus. Label the jars 7 and 8. Place one in a dark closet and leave one in the light. Examine both jars after several days.

2. Design experiments to show the effects of temperature and chemicals, such as table salt, on the growth of bread mold. Carry out the experiments under the teacher's supervision.

Chloroplast Pigment Analysis

When you look at a leaf, the green pigment, chlorophyll, is usually the only pigment that appears to be present. Actually, chlorophyll is only one of many types of pigments present in the leaf and one of several that are involved in the process of photosynthesis. Once removed from the leaf, the photosynthetic pigments can be separated from one another and identified using a process called chromatography.

Chromatography is a physical process in which several compounds are separated from a solution and from each other. In thin-layer chromatography, the solvent is absorbed by a thin layer of silica gel. As the solvent moves upward through the gel, it carries with it the compounds that have been placed on the gel. These compounds each move upward at a specific rate in relation to the moving solvent and can be identified by the distances they move.

OBJECTIVES

- Extract a mixture of plant photosynthetic pigments.
- Separate pigments of spinach leaves by thin-layer chromatography.

- Prepare and analyze a silica-gel chromatogram.
- Calculate the R_f values for various photosynthetic pigments.

MATERIALS

baby-food jar with lid
spinach leaves, dried
chromatography solvent
thin-layer chromatography slide
funnel
cheesecloth
dark-colored bottle or vial with stopper
10-mL graduated cylinder

capillary tube
metric ruler
pencil
ethyl alcohol
mortar and pestle
clock or watch
laboratory apron
goggles

PROCEDURE

Part A. Preparing For Chromatography

1. Obtain a small amount (approximately 5 mL) of the chromatography solvent from your teacher. Pour enough of this solvent into the baby-food jar so that it just covers the bottom of the jar but is less than 1 mm deep. Screw the lid onto the jar and set aside for later use.

2. Place a pea-sized amount of dried spinach in a mortar. Using the pestle, grind up the spinach for 2 minutes, as shown in Figure 1. Add 2 mL of ethyl alcohol to the ground spinach and continue to grind for another 2 minutes. The product should

Figure 1

LABORATORY MANUAL

Chloroplast Pigment Analysis

PROCEDURE continued

be a deep green fluid. Using a funnel, filter this fluid through a double layer of cheesecloth into a dark-colored bottle or vial. Stopper the bottle tightly until needed.

Part B. Making and Analyzing the Chromatogram

1. Select a chromatography slide. Handle it only by its edges. Make a small pencil dot 5 mm from the bottom of the slide. DO NOT use a pen to make the dot. Stand the slide next to the baby food jar with the dot at the bottom to verify that the dot is above the level of the solvent. If the solvent is too deep, pour a little of it out of the jar into a specially labeled container.

2. Dip a capillary tube into the pigment-containing fluid in the dark bottle.

3. Lightly touch the filled end of the capillary tube to the dot on the coated slide as shown in Figure 2. Allow a small amount of the fluid to be deposited on the slide, forming a spot 1 mm in diameter. Do not disturb the silica-gel film above the spot. Allow the spot to dry (about 30 seconds).

4. Repeat step 3, applying leaf pigments to the same spot four or five times, being sure to allow the spot to dry each time. This will produce a concentrated spot of pigments.

5. Place the baby-food jar on a level surface then place the slide inside of it, then place the slide inside of it, as shown in Figure 3. Do not allow the spot to contact the solvent at any time. Quickly screw the lid on the jar. Do not move the jar once the slide is placed in the solvent.

6. Watch the slide closely and note the movement of solvent up the film of silica gel. Remove the slide from the baby-food jar when the solvent front nearly reaches the top of the slide. Mark the top of the solvent front with a pencil as shown in Figure 4.

7. Make a drawing of your slide in the space provided in Data and Observations. Be sure to indicate the position of the original spot of pigments as well as the locations of pigments anywhere else on the slide. Indicate the relative amounts of pigments in each spot by drawing the spots the same size and darkness as those on your slide.

Figure 2

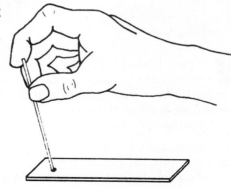

Figure 3

Figure 4

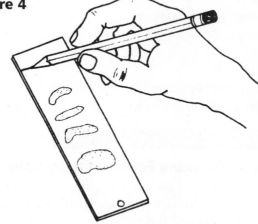

Carotenes, which are yellow or orange pigments, usually appear near the top of the slide. Lutein is a gray pigment just below the carotenes. Chlorophyll *a* will appear next as a blue-green pigment. Xanthophylls are yellow pigments, and chlorophyll *b* is a yellow-green pigment. They are found together just below

Chloroplast Pigment Analysis

chlorophyll *a*. Your chromatogram may or may not have all of these pigments.

8. Measure with a metric ruler the distance in mm from the original spot to the solvent front you marked in step 7. Record this measurement in Table 1.

9. Measure the distance each pigment traveled from the original spot to its final location. Record these data in Table 1.

10. Calculate the R_f value for each pigment spot. The R_f value is the ratio of the distance traveled by the pigment to the distance traveled by the solvent.

$$R_f = \frac{\text{distance pigment traveled}}{\text{distance solvent traveled}}$$

11. Record your R_f values as decimals rounded to the nearest hundredth in Table 1.

12. Clean your equipment and dispose of your solvents in the designated container. Wash your hands thoroughly with soap and water when finished with this lab.

DATA AND OBSERVATIONS

Table 1

Chromatography Data		
Substance	**Distance from original spot (mm)**	**R_f value**
Solvent front		_____
Carotenes		
Lutein		
Chlorophyll *a*		
Xanthophylls		
Chlorophyll *b*		

Your thin-layer slide

ANALYSIS

1. Which pigments were you able to identify?

_____

2. Judging from the darkness of the pigment spots on your chromatogram, which pigment would you say is most abundant in spinach leaves?

_____

3. Which pigment appeared to travel most rapidly? Least rapidly?

_____

4. Which pigment had the highest R_f value?

_____

5. How do the R_f values of the pigments compare with the rates of travel of the pigments?

_____

_____

Chloroplast Pigment Analysis

ANALYSIS continued

6. Why do the pigments travel in the solvent at different speeds? Remember that each pigment is a different molecule with its own characteristic size and mass.

_____

_____

_____

7. Do you think you would get similar results if you used a different kind of green leaf? Explain.

_____

_____

_____

8. Why do leaves appear green even though there are other pigments present?

_____

_____

_____

9. Many leaves change color in the autumn. How is it possible for this color change to occur? Base your answer on your new knowledge of pigments present in leaves. (HINT: Chlorophyll *a* and chlorophyll *b* are broken down in autumn when day length begins to shorten and temperatures decrease.)

_____

_____

_____

_____

FURTHER EXPLORATIONS

1. Conduct this Exploration using several different plants with differently colored leaves to see how their leaf pigments compare with those of spinach.

2. Separate pigments in spinach leaves by paper chromatography and compare the results with those obtained by thin-layer chromatography.

How Are Ferns Affected by Lack of Water?

The fern life cycle is said to show alternation of generations—a diploid sporophyte and a haploid gametophyte. Some stages of this life cycle are more dependent on water than others. In this Investigation, you will discover which stages occur under wet conditions and which ones occur under dry conditions.

OBJECTIVES

- Hypothesize how different stages of the life cycle of a fern are affected by water.
- Observe the release of spores from a fern plant.
- Observe the germination of fern spores.
- Observe the release of sperm from a fern prothallus.

MATERIALS

mature fern plant
scalpel
microscope slides (4)
plastic coverslips (4)
droppers (2)

glycerin
compound light microscope
wax marking pencil
toothpick
petroleum jelly

fern-spore culture
mature prothallus
pencil with eraser
water

laboratory apron
goggles

PROCEDURE

Part A. Alternation of Generations

1. Study the life cycle of the fern in Figure 1. Notice that the cycle includes two different forms of a fern: the large sporophyte fern plant and the small gametophyte called a prothallus. Locate these forms in the diagram. Notice that spores are produced by the leafy sporophyte fern plant.

2. Find the germinating spore in the diagram. A spore germinates to form a group of threadlike structures that develop into the tiny, heart-shaped prothallus. Notice that eggs and sperm are produced on the prothallus.

3. Make a **hypothesis** that states whether water is needed for each of these three stages of the fern life cycle: release of spores, germination of spores, and release of sperm. Write your hypothesis in the space provided.

Part B. Release of Spores

1. Obtain a fern plant. Look for brown dots on the back of the fronds of the fern. With a scalpel, carefully scrape one of these dots

onto a microscope slide. This structure contains several sporangia. **CAUTION:** *Use the scalpel with care. The tip and edge are sharp.*

2. Add a drop of glycerin then a coverslip to the slide. Glycerin will draw the water out of the sporangia. Tap the coverslip with the eraser end of a pencil.

3. Observe the slide under the low-power magnification of the microscope. When you find a sporangium, center it in the field of view and change to high-power.

4. Observe the sporangium for several minutes. Draw the sporangium before and after any change. Draw your observations of a sporangium in the space marked "Sporangium" in Data and Observations.

Part C. Germination of Spores

1. Use a wax marking pencil to write your initials on two microscope slides.

2. Use a toothpick to make a ring of petroleum jelly about the size of a dime on one slide. Place a drop of fern-spore culture inside the ring, and add a coverslip.

How Are Ferns Affected by Lack of Water?

Figure 1

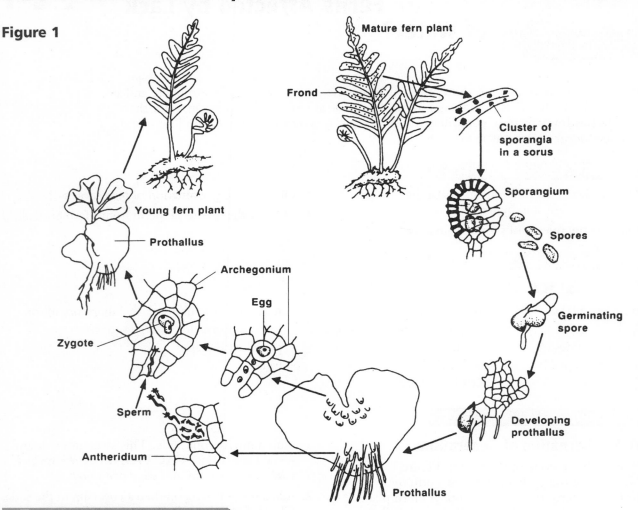

Mature fern plant

Frond

Cluster of
sporangia
in a sorus

Sporangium

Spores

Germinating
spore

Developing
prothallus

Prothallus

Young fern plant

Prothallus

Archegonium

Egg

Zygote

Sperm

Antheridium

PROCEDURE continued

3. Observe the slide under the low-power magnification of the microscope. Make a drawing of your observation in the space marked "Spore culture sealed slide" in Table 1.

4. Make a second wet mount of fern-spore culture, but do not use petroleum jelly to seal the coverslip to the slide.

5. Observe this slide under low-power magnification. Make a drawing of your observation in the space marked "Spore culture unsealed slide" in Table 1. Set the slide aside.

6. One or two days later, observe both slides again. Look for changes in the spores. Draw your observations in the appropriate spaces in Table 1.

Part D. Release of Sperm

1. Obtain a mature prothallus. Make a wet mount of the prothallus. Add a coverslip.

2. Observe the slide under low-power magnification. Near the notch in the prothallus, locate the structures that produce the eggs, called archegonia. Near the pointed base, locate the structures that produce the sperm, called antheridia.

3. Use the eraser end of a pencil to press gently on the coverslip. Watch closely to see if sperm are released. If sperm are released, observe their movement. Record your observations in Table 2.

4. Set the slide aside for several minutes, and allow it to dry out. Repeat step 3.

5. Clean all equipment as instructed by your teacher. Dispose of all items properly. Wash your hands thoroughly.

Lab
22-1

How Are Ferns Affected by Lack of Water?

HYPOTHESIS

_____

_____

DATA AND OBSERVATIONS

Sporangium

Table 1

Spore Culture Sealed Slide	Spore Culture Unsealed Slide
After 1–2 days	After 1–2 days

Table 2

Observations	Wet Prothallus	Dry Prothallus
Release of sperm		
Movement of sperm		

How Are Ferns Affected by Lack of Water?

ANALYSIS

1. Does adding glycerin to a slide simulate moist conditions or dry conditions? Explain you answer.

_____

_____

2. Describe what happened to the sporangium in glycerin. _____

_____

_____

3. Under what condition are spores released? _____

_____

4. How did the slides of spore cultures differ after a day or two? How did this difference affect the spores?

_____

_____

_____

5. Under which conditions were sperm released by the prothallus? _____

_____

_____

CHECKING YOUR HYPOTHESIS

Was your **hypothesis** supported by your data? Why or why not?

_____

_____

FURTHER INVESTIGATIONS

1. Obtain two fern plants of the same species and size. Plant them, using the same amount of soil for each. Keep the plants under the same conditions, varying only the amount of water they receive. Record the amount of water each plant receives. Also record the growth and development of each plant.

2. Investigate the conditions under which germinating fern spores grow best. Use some of the fern-spore culture from this Investigation. Determine the condition you would like to investigate, such as moisture, light, or growth medium. Design and set up an experiment to determine which conditions are best for the development of germinating spores.

LABORATORY MANUAL

EXPLORATION   Comparing Plants

Lab 22-2

A close examination of plant characteristics reveals a number of similarities and differences among plants. Vascular and nonvascular plants differ by the presence or absence of tissue made up of tubelike, elongated cells through which water, food, and other materials are transported. Seed plants and non-seed plants differ in their reproductive strategies. You will examine samples from three groups of plants: nonvascular plants—mosses, liverworts and hornworts; non-seed vascular plants—club mosses, horsetails, and ferns; and seed vascular plants—"naked" seed producers, which include cycads, ginkgoes, and conifers, and anthophytes, which have seeds enclosed in fruits.

OBJECTIVES

- Compare and contrast the physical features of vascular and nonvascular plants.

- Make a wet mount of a leaf from a vascular plant and a wet mount of a leaflike structure from a nonvascular plant for observation under a compound light microscope.

MATERIALS

hand lens
live, mature moss sample
live, mature leafy liverwort sample
live, mature horsetail sample
live, mature fern sample (sporophyte)
compound light microscope
scalpel
forceps

microscope slides (2)
coverslips (2)
dropper
water
live conifer with cones
live flowering plant with flowers or fruits
laboratory apron
goggles

PROCEDURE

Part A. Comparing Vascular and Nonvascular Plants

1. Use the hand lens to study the moss, liverwort, horsetail, and fern samples. Use Figure 1 to help you identify each sample.

2. Write the name of the division to which each sample belongs in Table 1. Use your textbook if necessary.

3. Observe closely the leaves or leaflike structures of each sample. Describe the size, thickness, and arrangement of the leaves/leaflike structures in Table 1.

4. Based on what you know about each plant division, do you think your samples are gametophytes or sporophytes? Write your answers in Table 1.

5. Use a hand lens to examine the plants for rhizoids, roots, and rhizomes. Rhizoids are small unicellular or multicellular strands of cells that help to anchor a plant. Roots are larger multicellular structures that anchor the plant to the ground and absorb water and nutrients from the soil, which are then transported to the stem by vascular tissue. Rhizomes are underground stems. Which of these structures does each of your samples have? Record your answers in Table 1.

6. Which plant samples are vascular plants? Which are nonvascular plants? Write your answers in Table 1.

Copyright © Glencoe/McGraw-Hill, a division of The McGraw-Hill Companies, Inc.

Comparing Plants

PROCEDURE continued

Part B. Using a Microscope to Examine the Leaflike/Leaf Structures of Mosses and Ferns

1. Using Figure 2 as a guide, use the forceps to remove one of the leaflike structures from the moss.

2. Place a drop or two of water on the center of a slide. Then place the leaflike structure in the water.

3. Place a second slide on top of the first. Press gently but firmly to crush the leaflike structure.

4. Remove the top slide. Place a cover slip over the crushed structure.

5. Examine the slide under the microscope. Is the leaflike structure one-cell thick or is it made up of layers of cells? Record your observations and make a sketch of what you see in Table 2.

6. With a scalpel, remove a tip from one of the fern fronds. **CAUTION:** *Use the scalpel with care. Cut in a direction away from your fingers.* The tip should be approximately the same size as the leaflike structure from the moss. Repeat steps 2–5 using the tip of the fern frond, and record your observations in Table 2.

Figure 1.

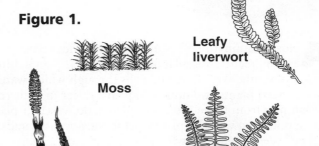

Leafy liverwort

Moss

Horsetail

Fern

Part C. Comparing Non-flowering Seed Plants and Anthophytes

1. Observe the structure of the conifer and the anthophyte. Use a hand lens to examine the leaves closely. Describe the shapes of the leaves in Table 3.

2. Are any reproductive structures present? Look for cones on the conifer. Look for flowers or fruits on the anthophyte. Complete Table 3 by illustrating or describing your observations. Be as detailed as possible.

Figure 2

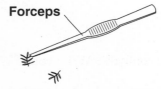

a. Remove leaflike structure from the moss.

Dropper

b. Place 1–2 drops of water on center of slide.

c. Place leaflike structure in center of water.

d. Place second slide on top of the first.

e. Press down gently but firmly to crush the leaflike structure.

f. Remove top slide and place a coverslip over the crushed leaflike structure.

**Lab
22-2**

Comparing Plants

DATA AND OBSERVATIONS

Table 1

Comparing Vascular and Nonvascular Plants					
Plant	Division	Descriptions of Leaves/Leaflike Structures	Gametophyte or Sporophyte?	Roots, Rhizoids, or Rhizomes?	Vascular or Nonvascular?
Moss					
Liverwort					
Horsetail					
Fern					

Table 2

Leaflike/Leaf Structures of Mosses and Ferns		
Plant	One-cell Thick or Layers of Cells?	Sketch
Moss		
Fern		

Table 3

Comparing Non-flowering Seed Plants and Anthophytes		
Plant	Leaves	Reproductive Structures
Conifer		
Flowering Plant		

Comparing Plants

ANALYSIS

1. Based on your observations, how are nonvascular plants different from vascular plants?

_____

_____

_____

2. How are rhizoids and roots similar? How are they different?

_____

_____

_____

3. How do you think nonvascular plants get water and nutrients from the soil? What about vascular plants?

_____

_____

_____

4. How does the thickness of the leaflike/leaf structures of mosses and ferns affect how water and nutrients move through the structures?

_____

_____

_____

5. How are conifers and anthophytes similar? How are they different?

_____

_____

_____

6. How might conifer leaves be better adapted for conserving water than the leaves of most anthophytes?

_____

_____

_____

FURTHER EXPLORATIONS

1. Visit a local park or wetland. Look for living examples of mosses, liverworts, horsetails, and ferns. Describe the environment in which each of the plants is found. If a park or wetland is not accessible, use nature photographs of the above plants.

2. Prepare short reports on *Welwitschia*, ginkgo, horsetails, yew, or the sago palm. Each of these plants has interesting historical or other information associated with it.

EXPLORATION   Roots and Stems

Plant roots and stems are living tissues that show a close relationship between their function and structure. Plant roots absorb water and minerals and have vascular tissues that transport them upward through the stem. Roots also store food and anchor plants.

One of the major functions of stems is support. Stems also are pathways for transporting food and water. Woody stems can be used to determine the ages of trees.

OBJECTIVES

• Identify and label root tissues.
• Identify the functions of root tissues.

• Identify and label the stems of one-, two-, and three-year-old trees.
• Describe the functions of stem tissues.

MATERIALS

cross section of   microscope slide   coverslip   laboratory apron
   parsnip root   iodine stain   compound light   goggles
single-edged razor blade dropper   microscope

PROCEDURE

Part A. Observation

The teacher has prepared cross sections of parsnip roots as shown in Figure 1.

1. Examine your cross section. You can see two different regions. These regions are marked A and B in Figure 1.

2. Prepare root tissue for microscopic viewing as shown in Figure 2 below.

A. Cut the root section in half. **CAUTION:** *The blade is sharp. Cut away from your fingers.*

B. Cut the root section in half again in the opposite direction to form a longitudinal section as shown.

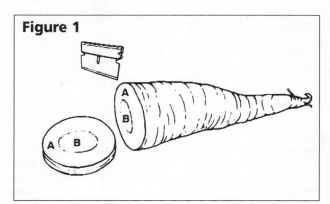

Figure 1

C. Slice off a thin section from one edge of the remaining root section.

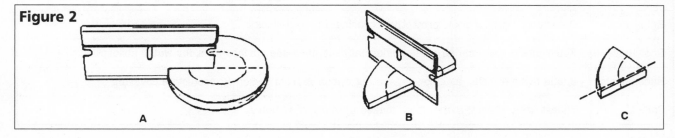

Figure 2

A   B   C

Roots and Stems

PROCEDURE continued

3. Place your thin section on a microscope slide. With a dropper, add iodine stain. Then cover with a coverslip. **CAUTION:** *Iodine is a poison. If spillage occurs, immediately wash with cold water and call the teacher.*

4. Examine the slide under the low-power magnification of the microscope. Slowly, move the slide across the field of view so that you observe all tissues of the root section. You will observe cells in region B shown in Figure 1 that look like railroad tracks. These are transporting cells. Cells in region A are rounded or squared and packed closely together in groups. These storage cells are much smaller than transporting cells. Starch grains in the storage cells will appear blue from the iodine stain.

5. Draw in Table 3 a few of the two different kinds of cells in regions A and B as seen under low power.

6. Complete Table 4 by giving the functions for the two regions. Note that the table gives the names for these regions.

7. Look at Figure 3, which shows a thin cross section of parsnip root seen under low-power magnification. This section was prepared as shown in Figure 4.

8. Identify and label the root structures in Figure 3 by using the descriptions in Table 1. Regions A and B (cortex and central cylinder) are labeled.

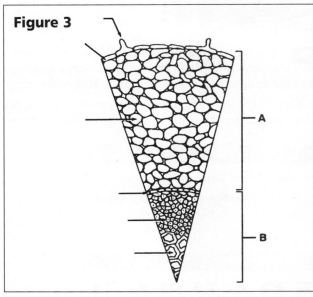

Figure 3

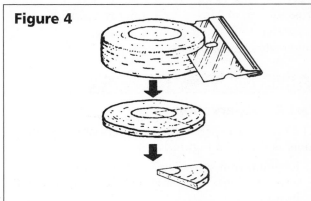

Figure 4

Table 1

Root Parts and Functions	
Tissue	**Description**
Xylem	Thick-walled cells In region B at tip end of slice; part of central cylinder; transports water
Phloem	Thin-walled cells in region B found in groups next to xylem; part of central cylinder; transports food
Epidermis	Outermost layer of root; protective covering; one cell thick
Root hairs	Fingerlike projections on some epidermal cells; increase surface area for water absorption
Endodermis	Single layer of cells, ringlike, separating region A from region B; protective covering
Cortex	Widest area of root; stores food; makes up most of region A

Lab
23-1

Roots and Stems

9. Clean all equipment as instructed by your teacher. Dispose of items properly. Wash your hands thoroughly when finished.

Part B. One-Year-Old Tree Stem

Figure 5 shows how a cross section of a tree stem is made.

1. Examine Figure 6, which shows the cross section of a one-year-old tree stem viewed through a microscope.

2. Study the descriptions of stem tissues in Table 2. Then label each kind of tissue in Figure 6.

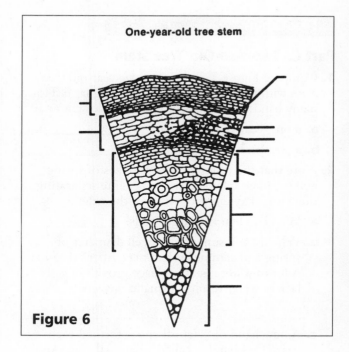

One-year-old tree stem

Figure 6

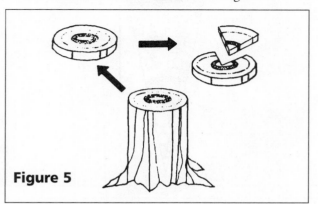

Figure 5

Table 2

Stem Tissues	
Tissue	**Description**
Cork	Outermost layer, about eight cell layers thick; protects against water loss
Cork cambium	Single layer of cells just inside the cork layer; produces new cork cells
Cortex	First layer inside cork cambium, about ten cells thick; cells larger and with thinner cell walls than cork cells; stores food
Pith	Tissue at center of stem (pointed end of wedge diagram); large thin-walled cells; stores food
Xylem	Thick layer of cells next to pith; widest layer of cells in stem; transports water and supports stem
(a) Spring xylem	Portion of xylem with large cells; produced in spring
(b) Summer xylem	Portion of xylem with small cells; produced in summer
Vascular cambium	Single layer of cells at outside edge of xylem; produces new xylem and phloem cells
Phloem	Groups of thin-walled cells, inside cortex; transports food
Bast fibers	Groups of thick-walled cells; surrounds phloem; supports stem

Roots and Stems

Part C. Two-Year-Old Tree Stem

1. Compare Figure 7, showing a two-year-old tree stem with the one-year-old stem in Figure 6. How many bands or sections of xylem are present in

a. a one-year-old tree stem? _____

b. a two-year-old tree stem? _____

2. Note that the difference in the sizes of spring and summer xylem cells forms a line separating one year of xylem growth from the other.

a. Which xylem has larger cells? _____

b. Why do you suppose that cell diameter of spring and summer xylem may differ? (HINT: A lot of water results in large growth of cells. A lack of water results in smaller growth.)

_____

c. A new band of xylem tissue is formed each year. This band is called an annual ring. An annual ring includes both spring and summer xylem. Which tissue forms this band?

_____

d. Each new band of xylem forms outside the older band of xylem. Which stem tissue is the oldest band of xylem found closest to?

_____

3. Label Figure 7, showing a two-year-old stem cross section. Use the following labels: *first-year xylem* (oldest), *second-year xylem* (youngest).

Part D. Three-Year-Old Tree Stem

1. Compare Figure 8, showing a three-year-old tree stem, with the one- and two-year-old stems.

2. Label the tissues in Figure 8. Use the label lines along the right side of Figure 8 to label each band of xylem as *first-year xylem* (oldest), *second-year xylem*, and *third-year xylem* (youngest).

3. Use the following terms to label the areas along the left side of Figure 8.

bark—all tissue from cork through vascular cambium
wood—all tissue from youngest xylem band through pith
vascular ray—narrow, one-cell-thick tissue extending through the xylem.

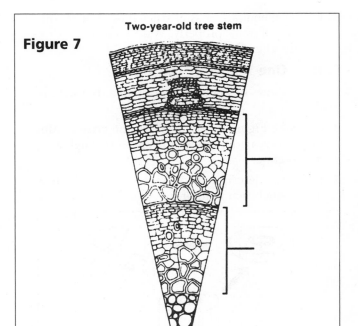

Figure 7 — Two-year-old tree stem

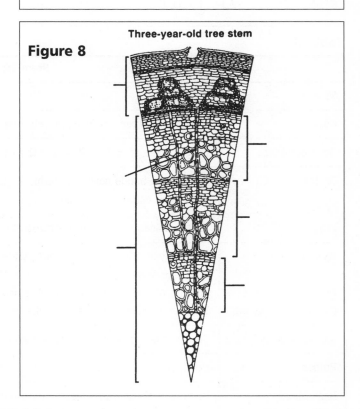

Figure 8 — Three-year-old tree stem

Roots and Stems

DATA AND OBSERVATIONS

Table 3

Parsnip Root Section Under Low-Power Magnification	
Region A	**Region B**

Table 4

Root Regions and Their Functions			
Region	**Name**	**Description**	**Function**
A	Cortex	Widest area of root	
B	Central cylinder	Center area of root	

ANALYSIS

1. How is the shape and function of the cells in the central cylinder of a root similar to a water pipe or blood vessel?

_____

_____

2. How is the structure of root cortex cells adapted for their function? _____

_____

3. Use Tables 1 and 2 to complete the following chart. Write the names of root and stem tissues that carry out each function listed.

Function	Roots	Stems
Protection		
Store food		
Transport food or water		
Absorb water		
Produce new tissue		
Support stem		

ANALYSIS continued

4. a. Name the main tissue types that make up the central cylinder of a root. _____

_____

b. What does each tissue type transport? _____

_____

5. Skin is called a dermis. The prefix *epi-* means outside. *Endo-* means inside.

a. What are the functions of epidermis and endodermis? _____

_____

b. Are these cell layers in roots properly named, based upon their location and function? Explain.

_____

_____

6. How is the structure of bast fibers adapted for their function? _____

_____

7. A thin, waxy layer is present along the outside of a tree's cork. Explain how this layer helps cork function.

_____

_____

8. a. How many bands of xylem does a three-year-old tree stem have? _____

_____

b. Does a new band of xylem form in a tree stem during each year of growth? _____

_____

c. How can a tree's age be determined? _____

_____

9. Annual rings vary in thickness due to environmental factors. What kinds of environmental factors during the year might influence the thickness of annual rings?

_____

_____

FURTHER EXPLORATIONS

1. Based on Figure 8 of a three-year-old tree stem, draw and label the cross section of a six-year-old tree stem, assuming similar environmental conditions.

2. Suppose that during a ten-year period, rainfall in a region increased each spring by ten percent. During the next ten-year period, rainfall decreased each spring by ten percent in the same region. What effect do you think this weather pattern would have on the annual rings of a 20-year-old tree that grew during those two ten-year periods? Construct a graph to illustrate your answer.

What Is the Effect of Light Intensity on Transpiration?

In order to carry out photosynthesis, a land plant must exchange gases with the atmosphere through its open stomata. One consequence of having open stomata is the loss of water vapor, called transpiration. Most of the water taken up by a plant's roots passes through the plant's vascular system and evaporates from the surfaces of its leaves during transpiration. Many different environmental factors, including temperature, humidity, light quality and intensity, and wind velocity, can influence a plant's transpiration rate.

OBJECTIVES

• Prepare a setup that will test the effect of light intensity on a plant's transpiration rate.

• Make a hypothesis to describe the effect of light intensity on the rate of transpiration.

• Compare the effects of high and low light intensities on transpiration rate.

MATERIALS

stopwatch
22-cm Pasteur pipette
permanent marker
pan, approximately 30 × 50 × 20 cm, with water approximately 10 cm deep
pruning shears
petroleum jelly
metric ruler

10-cm rubber surgical tubing, 5-mm bore
incandescent lamp
25- and 100-watt bulb (1 each)
dicot branch with leaves
thermal mitt
250-mL beaker
laboratory apron
goggles

PROCEDURE

1. With the permanent marker, make two marks on the pipette, the first mark 1 cm from the drip tip and the second mark 9 cm from the first mark. See Figure 1. **CAUTION: *The tip of the pipette is fragile. Do not press down on it.***

Figure 1

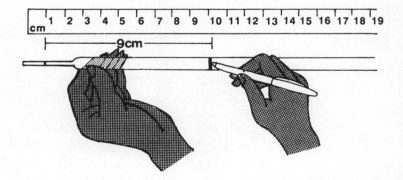

What Is the Effect of Light Intensity on Transpiration?

PROCEDURE continued

2. Spread a light coating of petroleum jelly around the wide end of the Pasteur pipette and around the dicot branch, 3 cm from the cut end. Be careful not to get petroleum jelly on the end of the branch. See Figure 2.

3. Slip the rubber tubing over the wide end of the pipette into the petroleum jelly. The fit should be snug with no air leaks.

4. Immerse the pipette and tubing in the pan of water so that the insides of both parts are completely filled with water.

5. Immerse the cut end of the dicot branch in the water. Use the pruning shears to cut off an additional 2 cm of stem under water. **CAUTION:** *Pruning shears are very sharp.* Slip the branch into the free end of the rubber tubing while keeping pipette, tubing, and branch under water. There should be no air bubbles in the setup. Compare your setup with that in Figure 3.

6. Test your setup for air bubbles by raising the drip tip of the pipette. Air bubbles except those immediately next to the branch will rise to the pipette tip. If there are air bubbles, disassemble and repeat steps 2 through 6.

7. Remove the pipette and branch from the water and place upright into a beaker. As the branch transpires, the water in the pipette will recede from the tip.

8. Make a **hypothesis** to describe the effect of light intensity on transpiration. Write your hypothesis in the space provided.

Figure 2

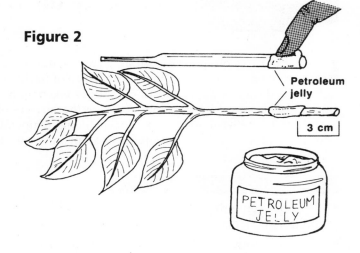

Petroleum
jelly

3 cm

PETROLEUM
JELLY

9. Position a lamp containing an incandescent 25-watt bulb so that the bulb is 15 cm away from the leaves of the branch. Turn on the lamp. Allow the plant to adjust under the light until the water in the pipette recedes to the 1-cm mark.

10. When the water reaches the 1-cm mark, start the stopwatch. When the water reaches the second mark, stop the stopwatch and record in the table under Trial 1 the time it took for the water to move 9 cm.

11. Refill the pipette by immersing the pipette tip in the pan of water. Gently squeeze the rubber tubing to expel the water and air. Release the pressure on the tube and refill the pipette with water.

12. Repeat step 7.

13. Repeat steps 9–12 two more times. Record your results in the table under Trials 2 and 3. Turn off the lamp.

Figure 3

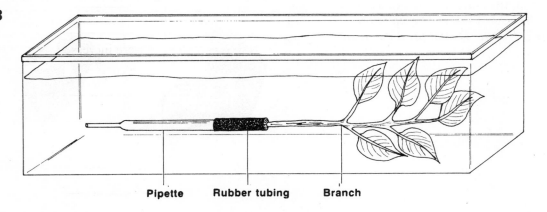

Pipette   Rubber tubing   Branch

Copyright © Glencoe/McGraw-Hill, a division of The McGraw-Hill Companies, Inc.

What Is the Effect of Light Intensity on Transpiration?

14. Change to the 100-watt bulb by disconnecting the lamp, then unscrewing the 25-watt bulb while wearing a thermal mitt. **CAUTION:** *The 25-watt bulb will be hot.* Repeat steps 9–13. Record your results in the table.

15. Calculate the average elapsed time for each light intensity and record these numbers in the table.

16. Clean all equipment as instructed by your teacher. Dispose of items properly. Wash your hands thoroughly.

HYPOTHESIS

_____

_____

DATA AND OBSERVATIONS

Table 1

Light Intensity	Elapsed Time (in seconds)			
	Trial 1	Trial 2	Trial 3	Average
Low (25-watt bulb)				
High (100-watt bulb)				

ANALYSIS

1. What does the movement of water in the pipette suggest about what is happening to water in the plant?

_____

_____

2. Through what part of the branch does water move? _____

_____

3. Compare the average time it took for water to move the 9 cm in the pipette at the two light intensities. At which intensity was the rate of movement greater?

_____

_____

4. Would you expect the trend you observed to continue if the light intensity were increased to that of a 200-watt bulb? Explain.

_____

_____

5. Why does a change in light intensity change the transpiration rate?

_____

_____

Lab

What Is the Effect of Light Intensity on Transpiration?

23-2

ANALYSIS continued

6. How would you expect the transpiration rate of a branch with twice as many leaves of the same size as your branch to compare with the rate you measured in this Investigation? Why?

_____

_____

7. In most plants, more stomata are on the lower surface of the leaves than on the upper surface. How is this advantageous to the plants?

_____

_____

CHECKING YOUR HYPOTHESIS

Was your **hypothesis** supported by your data? Why or why not?

_____

_____

FURTHER INVESTIGATIONS

1. Plan and conduct an experiment to test the effect on the transpiration rate of coating the upper and lower leaf surfaces with petroleum jelly.

2. Devise other experiments and hypotheses that test for the effects of wind, heat, humidity, or other environmental factors on the transpiration rate of plants.

Design Your Own

INVESTIGATION

How Do Environmental Factors Affect Seed Germination?

Seed plants produce seeds that nourish and protect the plant embryo. Seeds are produced in cones, pods, shells, or fleshy fruits, and they vary in size, shape, texture, and color. Many seeds require certain conditions in order to develop into a new plant, or germinate. Temperature, soil type, water, light, and pH are a few environmental factors that may influence a seed's ability to germinate. In this Investigation, you will design an experiment to test the effects of an environmental factor on the germination of seeds.

PROBLEM

How do different environmental factors influence seed germination? For example, will seeds germinate better in the dark or in direct sunlight? Which planting material allows the greatest number of seeds to germinate? How do temperature and moisture affect seed germination?

HYPOTHESES

Decide on one **hypothesis** that you will test. Your hypothesis might be, for example, that seeds need bright light to germinate or that seeds germinate more quickly at high temperatures. Write your hypothesis below.

_____

_____

OBJECTIVES

- Hypothesize how a particular environmental factor affects seed germination.

- Vary an environmental factor to observe its effect on seed germination.

- Compare seed germination under varying environmental conditions.

POSSIBLE MATERIALS

seeds (bean, radish, tomato, alfalfa)

planting material (potting soil, garden soil, sand, vermiculite, peat moss)

small, clear containers

light source (sunlight, fluorescent light)

paper towels

incubator

refrigerator

thermometer

water

acid solutions of various pH levels

basic solutions of various pH levels

pH indicator strips

marker

graph paper

plastic or rubber gloves

laboratory apron

goggles

Figure 1

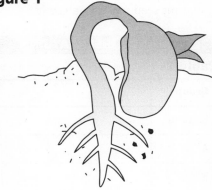

Germinating bean seed

How Do Environmental Factors Affect Seed Germination?

PLAN THE EXPERIMENT

1. Choose the type of seed you will use in your experiment. Decide on a procedure that will test the effect of the environmental factor you selected on seed germination. Use the suggested materials or others of your choosing. As you plan your procedure, think about what seeds need to germinate and what young plants need to grow.

2. Plant the seeds in containers. Label each container with the experimental condition to which those seeds will be exposed. For example, if you are testing the effect of temperature on seed germination, label each container with the temperature to which the seeds will be exposed. Describe in Table 1, or in a table of your own, the experimental condition to which the seeds in each container will be exposed.

3. Be sure you have a large enough sample of seeds in each container. Be aware, however, that the number of seeds you plant in each container may influence their germination.

4. Decide how and for how long you will collect data on the germination of the seeds. How will you know that a seed has germinated? You can record your data in Tables 2 and 3 or make your own tables.

5. Write your procedure on another sheet of paper or in your notebook. It should include the amount of each material you will need.

CHECK THE PLAN

1. Be sure that a control group is included in your experiment and that the experimental groups vary only in the environmental factors you are testing.

2. *Make sure the teacher has approved your experimental plan before you proceed further.*

3. Carry out your experiment. Wash your hands after working with the seeds and wear gloves when handling soils.

4. Check your seeds daily to see if any have germinated. Be sure to have a good working definition of germination so that you'll recognize a germinating seed when you see it.

Figure 2

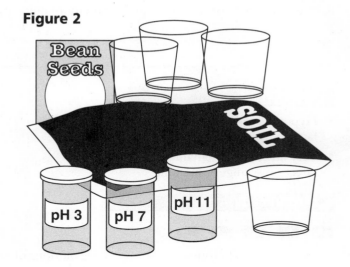

DATA AND OBSERVATIONS

Table 1

Group	Conditions to Which Seeds Are Exposed
Control	
A	
B	
C	
D	

How Do Environmental Factors Affect Seed Germination?

DATA AND OBSERVATIONS

Table 2

Date	Observations of Seeds				
	Control group	Group A	Group B	Group C	Group D

Table 3

Date	Number of Seeds Germinated				
	Control group	Group A	Group B	Group C	Group D

How Do Environmental Factors Affect Seed Germination?

ANALYSIS

1. Graph the data you collected on graph paper. Be sure to label the vertical and horizontal axes and to use the appropriate units for the axes. Based on your data, how does the environmental factor that you tested affect seed germination?

_____

_____

2. What variables did you keep constant in your experiment?

_____

_____

3. What were some possible sources of error in your experiment?

_____

_____

4. Exchange data with another group in your class that tested the same environmental factor you did. Do their data agree with yours? If not, provide possible reasons why.

_____

_____

5. Exchange data with a group in your class that tested a different environmental factor. What do their data indicate about the effect of that environmental factor on seed germination?

_____

_____

6. How could a nursery that is getting ready to prepare plants for spring planting utilize the information you obtained from your experiment?

_____

_____

CHECKING YOUR HYPOTHESIS

Was your **hypothesis** supported by your data? Why or why not?

_____

_____

FURTHER INVESTIGATIONS

1. Continue your experiment on seed germination to test the effects of the environmental factor you selected on the health of plants.

2. Repeat this Investigation using a different kind of seed. Compare your results with those from your first experiment. Explain how the data are similar and/or different. Give possible reasons for differences.

Lab

24-2

Do Dormant and Germinating Seeds Respire?

Dry seeds that are purchased for planting are alive but show no signs of any life processes. Scientists refer to seeds in this condition as being dormant. When dormant seeds are soaked in water, they respond by showing evidence of life processes. Seeds that show evidence of life processes are called germinating seeds. One of the life processes carried out by germinating seeds is cellular respiration. During respiration, a seed takes in oxygen from the air and gives off carbon dioxide. The volume of oxygen taken in and the volume of carbon dioxide given off are about equal.

OBJECTIVES

- Prepare respiration chambers to measure the amount of oxygen used by dormant and germinating seeds.

- Hypothesize whether dormant or germinating seeds use more oxygen.

- Compare the respiration rates of dormant and germinating seeds by measuring the amount of oxygen they use.

MATERIALS

test tubes (3)   metric ruler
bean seeds, soaked (5)   soda lime
bean seeds, dry (5)   small beaker
absorbent cotton   50-mL graduated
                                   cylinder

water, colored   rubber band
measuring half-teaspoon   laboratory apron
masking tape   goggles
marking pen

PROCEDURE

1. Place five dry, dormant bean seeds into a test tube.

2. Add a small cotton plug just above the seeds. Pack the cotton just tightly enough so that the seeds cannot fall out when the tube is turned upside down.

3. Add a half-teaspoon of soda lime (a carbon dioxide gas absorber) to the test tube. **CAUTION: *Do not handle soda lime with bare hands.***

4. Insert a second cotton plug over the soda lime as shown in Figure 1.

5. Use masking tape and a pen to label the test tube "Dormant." Place the label near the middle of the test tube.

6. Prepare a second test tube in a similar manner. However, put five soaked, germinating seeds in this test tube. Label this test tube "Germinating."

Figure 1

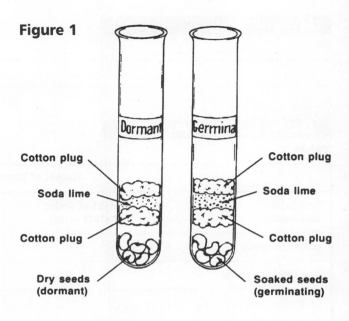

Cotton plug

Soda lime

Cotton plug

Dry seeds
(dormant)

Cotton plug

Soda lime

Cotton plug

Soaked seeds
(germinating)

Do Dormant and Germinating Seeds Respire?

PROCEDURE continued

7. Prepare a third test tube without seeds. Add a cotton plug, one half-teaspoon of soda lime, and a s second cotton plug just above the soda lime.

8. Place a rubber band around all three test tubes to prevent them from tipping over. Turn all the test tubes upside down and place them into a small beaker containing 30 mL of colored water.

9. With a metric ruler, measure in millimeters the height to which the colored water rises in each test tube (Figure 2). Measure the water level in the test tube, not in the beaker.

10. Record in Table 1 the height to which the colored water rises in each test tube. Record the height as 0 mm if water does not move into a test tube.

11. Do not disturb the test tubes for at least 24 hours.

12. Make a **hypothesis** to explain which one of the test tubes will show the greatest rise in water level. Write your hypothesis in the space provided.

13. After 24 hours, measure and record the height of water inside each test tube. DO NOT move the beaker or remove the test tubes from the beaker until you have measured the height of water within each test tube.

Figure 2

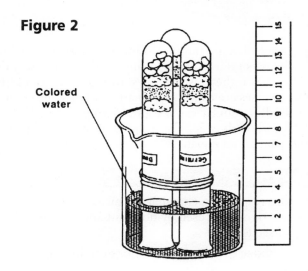

Colored water

14. Calculate the rise in water level from start to end of the Investigation by subtracting the height of water in each tube at the start from the height of water in each tube after 24 hours. Record this amount in the last column of Table 1.

HYPOTHESIS

_____

_____

DATA AND OBSERVATIONS

Table 1

Height of Water in Test Tubes			
Test tube contents	Height of water at start (mm)	Height of water after 24 hours (mm)	Rise in water level (mm)
Dormant seeds			
Germinating seeds			
No seeds			

Do Dormant and Germinating Seeds Respire?

Lab 24-2

ANALYSIS

1. a. What do you think is trapped within each test tube turned upside down in the water?

_____

b. What prevents water from moving into each test tube at the start of the experiment?

_____

2. a. Which test tube shows the most air used in 24 hours?

_____

b. What is the evidence?

_____

c. What gas in the air is being used?

_____

3. What life process is responsible for the use of this gas?

_____

_____

4. a. Which kind of seed, dormant or germinating, carries out this process at a faster rate?

_____

b. What is the evidence?

_____

_____

_____

5. What does the test tube without seeds show?

_____

_____

_____

6. Soda lime absorbs carbon dioxide from air.
a. Recalling that seeds use oxygen and release carbon dioxide, predict the level of water in the test tube with germinating seeds after 24 hours if soda lime is not used.

_____

b. Explain.

_____

_____

Do Dormant and Germinating Seeds Respire?

ANALYSIS continued

7. A seed contains a food supply. This food is used during respiration.

a. Which type of seed, dormant or germinating, would use food at a slower rate?

_____

b. What evidence do you have to support your answer?

_____

_____

c. Explain why using food slowly may help plants survive.

_____

_____

8. Suppose you place 20 germinating bean seeds into one test tube and add soda lime. Into a second test tube, you place 10 germinating bean seeds and add soda lime. You then invert both test tubes into a beaker of water. Explain what will happen.

_____

_____

_____

9. Suppose you boil some germinating seeds for 30 minutes and place these boiled seeds into a test tube with soda lime. You then prepare a second test tube with unboiled germinating seeds and soda lime and invert both test tubes into water. Explain what will happen.

_____

_____

_____

CHECKING YOUR HYPOTHESIS

Was your **hypothesis** supported by the data you recorded? Why or why not?

_____

_____

FURTHER INVESTIGATIONS

1. Try the same Investigation using other kinds of seeds, such as grass, radish, or corn. Do you think that you will get similar results?

2. Do differences in water temperature affect the respiration of dormant and germinating seeds? Design an experiment that will show the effect of hot water (32°C), warm water (22°C), cool water (12°C), and cold water (2°C) on the respiration of dormant and germinating seeds.

EXPLORATION Symmetry

Animals vary in their patterns of embryonic development, type of support, mechanisms for food capture, type of nervous response, structure of digestive and transport systems, methods of excreting wastes, and methods of locomotion. Animals also exhibit differences in symmetry, or the arrangement of their body parts. Organisms that lack a regular arrangement of parts are categorized as asymmetric. Animals with parts that radiate from a central point or from a central axis have radial symmetry. Animals that can be divided into similar halves along a central plane are bilaterally symmetrical. Animals with bilateral symmetry have definite front and back (ventral and dorsal) areas and a concentration of nervous tissue in the anterior region, often with external sense organs. The symmetry of an animal affects the way it moves and obtains food.

OBJECTIVES

- Identify the symmetry of a variety of animals.
- Relate symmetry to methods of food capture and locomotion.

MATERIALS

preserved or dried
  specimens of a sand
  dollar, sea urchin,
  brittle star, jellyfish,
  butterfly or moth,
  sea anemone, and
  squid
live crayfish, goldfish,
  lizards, grasshoppers

living cultures of
  Daphnia, planarians,
  Hydra, rotifers
yeast boiled in
  Congo red
beef liver
living flies
lettuce
guinea pig food
pipettes (3)

fish food
aquarium or fish bowl
terrarium with
  screened top
spring, pond, or
  aquarium water
watchglass
compound light
  microscope

10-mL graduated
  cylinder
dissecting microscope
depression slide
coverslip
glass jars, large (2)
plastic or rubber gloves
laboratory apron
goggles

PROCEDURE

Part A. Identifying Symmetry

1. Study the animals shown in Figure 1. Identify the type of symmetry that characterizes each animal's body plan and record it in Table 1. Note whether the organism can be divided along any plane into similar halves. If so, classify the organism as radially symmetrical. If the organism is irregular in shape, then its body plan exhibits asymmetry. Can the organism be divided along its length to form similar halves? Does the anterior end look different from the posterior end? Does the dorsal surface look different from the ventral surface? Then the organism is bilaterally symmetrical.

2. Examine each of the preserved or dried specimens and record their symmetry in Table 2.
   CAUTION: *Wear protective gloves when handling preserved specimens.*

Part B. Symmetry and Behavior

In this part of the Exploration, you will be working with a variety of living organisms to identify their symmetry and observe how it affects the way they move and feed. Follow your teacher's instructions regarding the care and handling of the organisms. Use your observations and information from your text to record the symmetry of each animal in Table 3.

Symmetry

 PROCEDURE continued

1. **Planarians:** Place a few drops of water on a watchglass. Use a pipette to remove a planarian from the culture jar and place it in the water. Introduce a small piece of liver, and observe the movement and feeding behavior of the planarian by using a dissecting microscope. Record your observations in Table 3.

2. **Rotifer:** Using a pipette, place a drop of the rotifer culture on a depression slide. Add a drop of the stained yeast to the slide and place a coverslip over it. Use a compound light microscope to examine the rotifers' method of obtaining and ingesting the yeast cells. Record your observations.

Figure 1

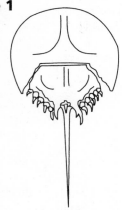

horseshoe crab

turtle

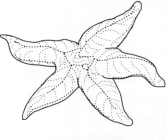

sea star

skate

sponges

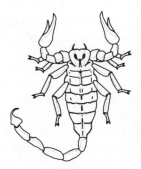

dog

scorpion

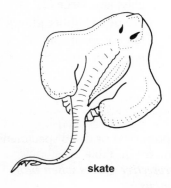

millipede

Symmetry

3. _Hydra:_ Place some water on the watchglass. Use a pipette to add one or more _Hydra_, and then place the watchglass under the dissecting microscope. Release about 0.5 mL of the _Daphnia_ culture near the _Hydra_. Observe the _Hydra_ under the microscope as they capture and eat the _Daphnia_. Record your observations in Table 3.

4. Crayfish: Place a crayfish in an aquarium or fishbowl and fill it with water. Drop a few pellets of guinea pig food a short distance in front of the crayfish. If it is hungry, it should respond quickly to the food. Record its movements and feeding behavior in Table 3.

5. Goldfish: Place a goldfish in a large jar of aquarium water. Sprinkle some flakes of food on the surface of the water. Observe and record the behavior of the fish.

6. Lizard: The lizard can be observed in a terrarium that has a screened top. Living flies are the food source. Do not distract the lizard after introducing the flies into the terrarium. Watch patiently for food capture and consumption.

7. Grasshopper: Observe the grasshopper's movements in a large glass jar. Slowly place a small piece of lettuce in the jar and observe how the insect responds. Avoid distracting the grasshopper with your own movements.

DATA AND OBSERVATIONS

Table 1

Organism	Symmetry
Horseshoe crab	
Sea star	
Turtle	
Skate	
Dog	
Sponge	
Millipede	
Scorpion	

Table 2

Organism	Symmetry
Sand dollar	
Sea urchin	
Brittle star	
Jellyfish	
Sea anemone	
Butterfly/moth	
Squid	

Table 3

Organism	Symmetry	Description of Movement and Eating Behavior
Planarian		
Rotifer		
Hydra		
Crayfish		
Goldfish		
Lizard		
Grasshopper		

Symmetry

ANALYSIS

1. What do your observations suggest about the relationship between an animal's symmetry and its method of movement?

_____

_____

_____

_____

2. What do your observations suggest about the relationship between an animal's symmetry and its method of food capture?

_____

_____

_____

_____

_____

3. What are the advantages and disadvantages of radial symmetry?

_____

_____

_____

_____

4. What appears to be the evolutionary trend in symmetry? Why do you think this is so?

_____

_____

_____

_____

_____

FURTHER EXPLORATIONS

1. Design an investigation that would compare the internal and external symmetry of an animal.

2. Using the library and other textbooks as resources, show the correlation between early embryonic development and the symmetry of at least three organisms.

EXPLORATION Hydra Behavior

Hydra are usually sessile organisms. They capture prey with their nematocysts, located on their tentacles. Their nerve net controls the movement of the tentacles as the prey is brought to the mouth and into the gastrovascular cavity. In this Exploration, you will observe the triggering of nematocysts and determine the kinds of stimuli that cause the triggering of nematocysts.

OBJECTIVES

- Prepare a wet mount of a hydra for microscopic observation .
- Observe the triggering of nematocysts caused by dilute acid.
- Observe the reactions of a hydra to food and non-food stimuli.

MATERIALS

droppers (3)   *Hydra* culture
depression slides (2)   dissecting microscope
dilute acetic acid   small beaker
aquarium water   dissecting probe
Daphnia culture   laboratory apron
                                  goggles

PROCEDURE

1. Using a dropper, transfer a hydra and a few drops of culture water to a depression slide. Allow time for the hydra to relax and extend its tentacles. **CAUTION: *Hydra are living animals. Handle them with care.***

2. Observe the hydra under low and then high power of the microscope. See Figure 1.

3. With the hydra in view, carefully add one drop of dilute acetic acid to the depression slide. **CAUTION: *If acetic acid is spilled, rinse with water and call your teacher immediately.*** The acetic acid will cause the nematocysts to discharge. If

they do not discharge immediately, carefully add acetic acid one drop at a time until they do.

4. After observing the triggering of nematocysts, remove the hydra to a small beaker of aquarium water by dipping the slide into the beaker.

5. Using a dropper, transfer a few hydra and a few drops of culture water to the second depression slide. Allow the hydra to relax and extend their tentacles.

6. With a hydra in view under the high power of the microscope, carefully shake the depression slide. In Table 1, record the reaction of the

Figure 1

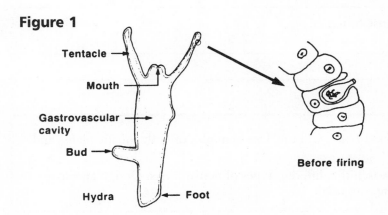

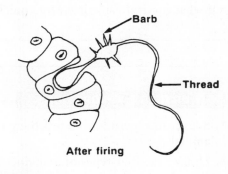

Before firing

After firing

Nematocyst

Copyright © Glencoe/McGraw-Hill, a division of The McGraw-Hill Companies, Inc.

Hydra Behavior

PROCEDURE continued

hydra. With the dissecting probe, carefully touch the body and a tentacle of the hydra. Record the reaction of the hydra.

7. Allow the hydra to relax. Then, use a dropper to transfer a few *Daphnia* to the depression slide. In Table 1, record the reaction of the hydra when a *Daphnia* comes close to it.

DATA AND OBSERVATIONS

Table 1

Reactions of *Hydra*	
Stimulus	**Response**
Shaking of slide	
Touching with probe	
Presence of *Daphnia*	

ANALYSIS

1. Why do you think the nematocysts trigger when *Hydra* is exposed to acetic acid?

_____

2. How does the reaction of *Hydra* to *Daphnia* differ from its reaction to being shaken or touched?

_____

_____

3. Describe the feeding process of *Hydra*.

_____

_____

4. *Hydra* cannot pursue their prey. What adaptation makes this unnecessary?

_____

5. Of what survival value is *Hydra's* quick response to touch?

_____

_____

FURTHER EXPLORATIONS

1. Set up an experiment to test the types of food *Hydra* will eat. Use examples of both aquatic plants and animals.

2. *Hydra* have four types of nematocysts. Research the different types of nematocysts. Design an experiment to observe the four types and how they are used.

Planarian Behavior

Planarians are free-living flatworms that usually measure less than one centimeter in length. The head area of the planarian contains a brainlike concentration of nerve cells. The nerves and digestive system form a ladderlike arrangement within the flattened body. In this Exploration, you will observe how planarians move, how they react to touch, and their feeding behavior.

OBJECTIVES

- Observe how a planarian moves using both muscles and cilia.
- Observe the feeding behavior of a planarian.
- Observe how a planarian reacts to touch and to environmental disruption.

MATERIALS

dropper
aquarium water
hand lens
watchglass
minced liver

live planarians (3–4)
petri dish
dissecting microscope
dissecting probe
laboratory apron
goggles

PROCEDURE

Part A. Movement and Feeding

1. Using a dropper, carefully transfer at least two planarians to a watchglass. **CAUTION: *Handle planarians with care*.** Add enough aquarium water to cover the bottom. Planarians should be able to move freely.

2. Place the watchglass on the stage of the microscope. Allow a few seconds for the planarians to recover from being handled.

3. Using Figure 1 as a reference, locate the mouth and pharynx in the middle of the planarian's underside.

4. Observe how the planarians move. A planarian has two layers of muscles and a lower surface covered with cilia. Can you see the wavelike motion of the cilia? How would you characterize the movement of planarians? Record your observations in Table 1.

5. Place a speck of minced liver in the watchglass with the planarians. Observe the feeding behavior. Look for the pharynx as it extends through the mouth. Can you see the intestines? What position is the planarian in while it feeds? How long does feeding take? Record your observations in Table 1.

Part B. Response to Touch

1. In a petri dish, transfer at least two planarians with the dropper. Add enough aquarium water to just cover the planarians. Allow the planarians some time to relax.

2. With the tip of the probe, gently tap on the side of the petri dish. How do the planarians react? Record your observations in Table 2. Wait at least one minute. Gently touch the head of one of the planarians with the side of the probe. How does it react? Gently touch the tail in the same manner. How does it react?

Figure 1

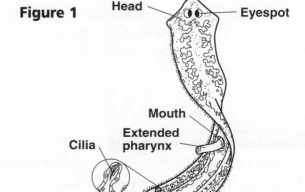

Head
Eyespot
Mouth
Extended pharynx
Cilia
Cilia

Planarian Behavior

DATA AND OBSERVATIONS

Table 1

Movement and Feeding Behavior of Planarians	
Behavior	**Observations**
Movement	
Feeding	

Table 2

Reactions of Planarians to Touch and Environmental Movement	
Stimulus Response	
Tapping the petri dish	
Touching the head	
Touching the tail	

ANALYSIS

1. Describe the planarian's behavior once the liver was placed in the watchglass.

_____

_____

2. Consider the entire digestive process of the planarian. What do you think is the adaptive advantage to having the planarian's mouth located in the center of the body?

_____

_____

_____

3. How is the flat shape of the planarian adaptive to its feeding method?

_____

_____

4. What part of the body first reacts to light, touch, or other changes in the environment? How do your observations support this?

_____

_____

FURTHER EXPLORATIONS

1. Set up a similar exploration to observe the reaction of planarians to sound.

2. Design an experiment to observe the long-term feeding behavior of planarians. Determine feeding intervals and quantities for planarians.

Squid Dissection

Mollusks are soft-bodied animals and include snails, clams, octopuses, squids, and slugs. They are bilaterally symmetrical and have well-developed digestive, circulatory, excretory, and respiratory systems. The class Cephalopoda, which includes the squid and the octopus, is considered to be the most complex and highly developed group of mollusks. Squids are characterized by a large prominent head with conspicuous eyes and a mouth surrounded by ten tentacles.

OBJECTIVES

- Dissect a squid and identify the organs and major organ systems of the squid.
- Determine the functions of various squid structures.
- Describe features of the squid that are characteristic of phylum Mollusca.

MATERIALS

squid (fresh or thawed)   stereomicroscope or
scissors   hand lens
dissecting pins (5–10)   laboratory apron
dissecting pan   goggles

PROCEDURE

Part A. Body Organization

1. Put on the laboratory apron. Place the squid in the dissecting pan. Note that there is no external shell and that the major part of the body is enclosed by the soft, muscular **mantle**. There are ten conspicuous arms, or **tentacles**, derived from the mollusk foot.

2. Arrange your specimen so that the mantle points away from you. See Figure 1. Turn the animal so that the **siphon** faces you. The eyes should be on the right and left sides of the body.

3. Slit open the mantle cavity by inserting the tip of the scissors under the mantle at the siphon and cutting to the pointed end of the mantle. Cut with care so that you do not disturb the internal organs.

4. Pin down the mantle to the pan, slanting the pins at an angle away from the specimen as shown in Figure 2.

Figure 1

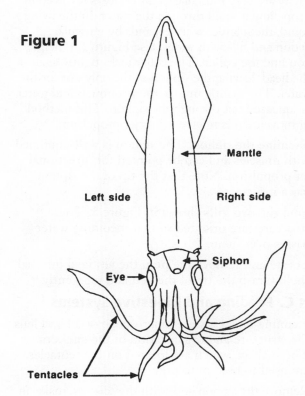

Mantle

Left side   Right side

Siphon

Eye

Tentacles

Squid Dissection

Figure 2

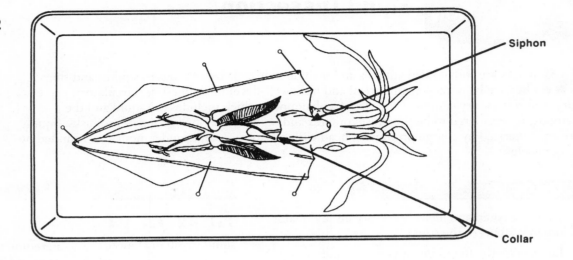

Siphon

Collar

Part B. Mantle Cavity and Respiratory System

1. Examine the mantle cavity. The walls of the mantle cavity are very muscular. This cavity is involved in propelling a squid through the water. In the living squid, the mantle cavity expands by muscular action and fills with water. Use Figure 2 to help you find the **collar**. The collar locks tightly against the head, leaving the siphon as the only exit for the water. The mantle muscle then contracts and water is squeezed out through the siphon. This method of movement is referred to as jet propulsion.

2. Examine the siphon. The siphon is well-equipped with muscles and can be pointed for directional jet propulsion. Note that the tip of the siphon has a muscular valve.

3. Find the two **gills** shown in Figure 3. These structures are oriented so that incoming water passes over them.

4. Locate and remove the **pen**, the vestigial internal shell. Grasp the tip of the pen and tug gently.

Part C. Feeding and Digestive Systems

1. Examine under a stereomicroscope or a hand lens the structure and organization of the **suckers**. The suckers, which are located on the tentacles, are used to hold onto prey.

2. Remove the siphon and, with the scissors, make an incision into the head. Expose the beak as shown in

Figure 3

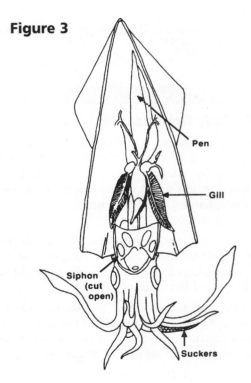

Pen

Gill

Siphon (cut open)

Suckers

Figure 4. Pry open the beak and observe the tonguelike **radula**. Trace the **esophagus**, which is surrounded by the liver, to the thick-walled **stomach**. The stomach emerges to form the **caecum**. Note the **pancreas**. The **intestine** runs from the stomach and terminates in the **rectum**. An **ink sac** arises from the intestine near the anus. The ink sac is used for defense.

Squid Dissection

Figure 4

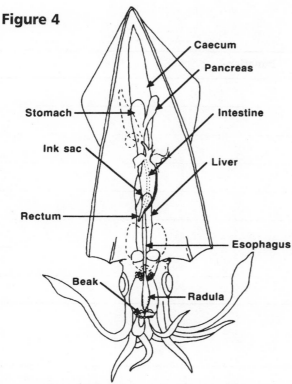

Figure 5

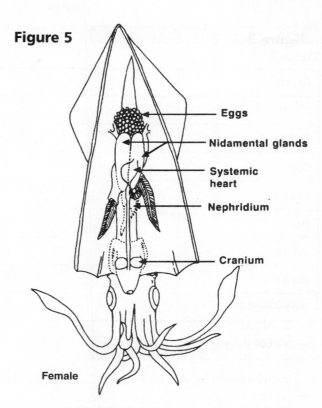

Female

Part D. Circulatory, Excretory, Nervous, and Reproductive Systems

1. Locate the **systemic heart** as shown in Figure 5. This is a difficult structure to find because it is transparent.

2. Examine the **nephridium**, a kidneylike excretory organ that removes waste products from the blood.

3. Locate the white mass of the **cranium** above and between the eyes. This structure contains the squid's **brain**.

4. Locate the reproductive organs as shown in Figures 5 and 6. Determine the sex of your squid, but be sure to examine squids of both sexes. The male has **testes** that lie beneath the caecum. The female has two large **nidamental glands** that secrete a protective covering over the **eggs**. Eggs might or might not be present.

5. Remove the **eye** and cut it in half. Examine the transparent **lens** and the shiny black **retina** at the back of the eye.

6. Complete Table 1 in Data and Observations. Consult your textbook if necessary.

7. Wash your hands thoroughly with soap and water after handling the squid.

Figure 6

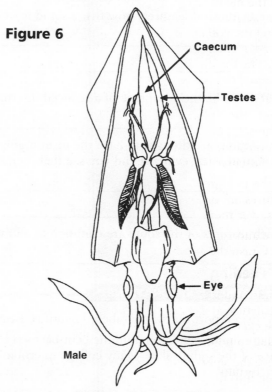

Male

Squid Dissection

DATA AND OBSERVATIONS

Table 1

Organ	Function
Mantle	
Siphon	
Gills	
Suckers	
Pen	
Ink sac	
Nephridium	
Nidamental gland	

ANALYSIS

1. Describe squid locomotion. _____

_____

2. What anatomical features show that a squid is well-adapted to a predatory existence?

_____

_____

3. What could be the advantage of a reduced internal shell? _____

_____

4. Cephalopods are thought to have the most highly developed eyes in the invertebrate world. What anatomical evidence did you see that indicates this?

_____

_____

5. Which features of the squid are common to all members of the phylum Mollusca?

_____

FURTHER EXPLORATIONS

1. Write a report on the chambered nautilus. Learn how it controls its depth in the water.

2. Many squids contain bioluminescent bacteria in their skin. Try to culture *Photobacterium* from the surface of the squid's skin. Many college microbiology lab books contain directions on how to culture this bacterium.

INVESTIGATION

How Do Snails Respond to Stimuli?

The organization of an animal's nervous system often reflects its patterns of behavior. Animals that are very active usually show a greater degree of brain development and have better-developed sense organs than animals that are not very active. Though not a rapid mover, the snail is an active animal and its nervous system is well-developed. The snail has sense organs in its mantle cavity and at the ends of its tentacles. The way it responds to environmental stimuli may mean the difference between survival and death.

OBJECTIVES

- Observe a snail's response to gravity, touch, and light.
- Test a snail's response to a chemical substance and to substances common in its environment.

- Make a hypothesis that describes a snail's response to an acid substance.
- Measure the speed at which a snail moves.

MATERIALS

white paper
stereomicroscope or hand lens
lettuce
pencil
vinegar or lemon juice

glass jar (1 pint) or 250-mL beaker
paper towels
live snails (5)
moist soil
camel-hair brush

small plastic tray with lid, approximately 15 cm × 20 cm × 5 cm
cardboard, approximately half the size of the tray

metric ruler
stopwatch or clock with second hand
water
laboratory apron
goggles

PROCEDURE

Part A. Geotaxis and Touch

Geotaxis is the response of an animal to the force of gravity. The response may be either positive (toward gravity) or negative (away from gravity).

1. Place five snails on a wet paper towel on a flat surface.

2. Locate the head region and tentacles of the snails. Using a hand lens or a stereomicroscope, locate the eyes on the ends of the tentacles as shown in Figure 1.

3. Place two or three snails in the bottom of a clean jar or beaker. Place the jar on its side and observe snail movement. Return the jar to an upright position and place the snails on the bottom again. Place the jar on its side and observe. Record in Table 1 how many snails moved toward the force of gravity and how many moved away from the force of gravity.

Figure 1

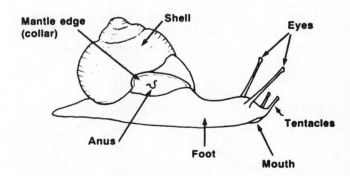

4. Very gently tickle the tentacles with the paintbrush and observe the behavior of the snails. Record your observations in Table 1.

How Do Snails Respond to Stimuli?

PROCEDURE continued

Part B. Phototaxis

Phototaxis is the movement of an animal in response to light. The phototactic response can be either positive (toward light) or negative (away from light).

1. Place a piece of paper towel in the bottom of a small plastic tray.

2. Divide the tray into two halves by marking a line on the paper towel with a pencil as shown in Figure 2.

3. Moisten the paper towel with water and place the five snails on the line.

4. Place a piece of cardboard on the top of one-half of the tray. This will provide a darkened area for the snails.

5. Observe the snails for 15 minutes.

6. Record in Table 1 how many snails showed a positive phototactic response and how many snails showed a negative response.

Part C. Chemotaxis

Chemotaxis is the movement an animal makes in response to a chemical stimulus. Chemotactic movements can be either positive (toward the chemical) or negative (away from the chemical).

1. Remove the moist paper towel from the plastic tray.

2. Divide the plastic tray into halves with a pencil line.

3. Place a piece of lettuce in one half of the tray as shown in Figure 3.

4. Place the five snails on the center line and cover the tray with its lid. Observe the activity of the snails for 15 minutes. Record your observations in Table 1.

5. Repeat steps 3 and 4, substituting moist soil for the lettuce.

6. Repeat step 3, substituting a paper towel soaked in vinegar or lemon juice for the lettuce.

7. Make a **hypothesis** that describes how the snails will react to this acid substance. Write your hypothesis in the space provided.

8. Repeat step 4.

Part D. Snail's Pace

1. Draw a short line on a piece of white paper.

2. Moisten the paper with water.

3. Place the front of one snail on the line and allow the snail to move for 60 seconds.

4. Measure the distance (in mm) traveled by the snail. Record the distance in Table 2 for Trial 1.

5. Repeat steps 3 and 4 three more times. Calculate the average speed first in mm per 60 seconds (mm/60 s) and then in mm per second (mm/s). Record your results next to Table 2.

6. Compare the speed of your snail with those of others in your class.

7. Return the snails to their proper place, as directed by your teacher, at the conclusion of this Investigation. Wash your hands thoroughly with soap and water after handling the snails.

HYPOTHESIS

_____

_____

_____

_____

Figure 2

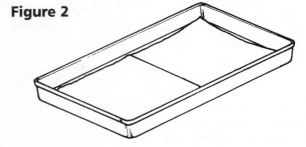

Figure 3

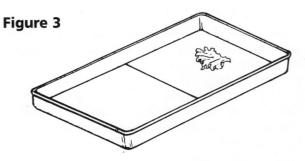

How Do Snails Respond to Stimuli?

DATA AND OBSERVATIONS

Table 1

Responses of Snails to Stimuli	
Test	Response
Geotaxis	
Touch	
Phototaxis	
Chemotaxis Lettuce	
Soil	
Vinegar or lemon juice	

Table 2

A Snail's Pace (mm)	
Trial 1	
Trial 2	
Trial 3	
Trial 4	

average speed = _____ mm/60 s

average speed = _____ mm/s

ANALYSIS

1. a. Was the snails' behavior toward or away from the force of gravity?

_____

b. Is this positive or negative geotaxis?

_____

2. a. How did the snails react when given the choice between light and dark conditions?

_____

b. How might this type of behavior be beneficial to a snail?

_____

_____

Lab 27-2

How Do Snails Respond to Stimuli?

ANALYSIS continued

3. How did the snails react when touched on the tentacle?

_____

4. a. How did the snails react when exposed to moist soil and lettuce?

_____

_____

b. Explain how this behavior could be beneficial.

_____

_____

5. a. Did the snails exhibit a positive or a negative response to the vinegar or lemon juice?

_____

_____

b. How might this type of behavior be beneficial to a snail?

_____

_____

6. Although a snail moves slowly, why is it considered to be an active animal?

_____

_____

_____

CHECKING YOUR HYPOTHESIS

Was your **hypothesis** supported by your data? Why or why not?

_____

_____

_____

FURTHER INVESTIGATIONS

1. Design an experiment to test a squid's response to the same stimuli. Compare the squid's responses with those of the snail. Account for any differences in response by comparing the environments of the animals.

2. Squids have chromatophores in their skin, which contain various colored pigments. Chromatophores are under nervous and hormonal control and can create a complex pattern of colors and color flashes by turning on and off. The color patterns created are used in communication. Research this subject in the library and prepare a report.

EXPLORATION   ## Comparing Arthropods

Classification involves examining the characteristics of organisms and placing the organisms into groups based on their shared characteristics. At the same time, however, the organisms in a group may have many differences. Phylum Arthropoda has more species than any other group of animals. Entomologists suggest that there may be over 800,000 arthropod species. Members of this phylum share certain characteristics, but within the group, three major classes have been identified—the crustaceans, the insects, and the arachnids—based on differences among the animals.

OBJECTIVES

- Identify the characteristics of a spider, a crayfish, and a grasshopper.
- Determine which characteristics are phylum characteristics.
- Determine which characteristics are class characteristics.
- Construct a dichotomous taxonomic key.

MATERIALS

preserved spider
preserved crayfish
preserved grasshopper
dissecting probe

dissecting pan
stereomicroscope or hand lens
plastic or rubber gloves
laboratory apron
goggles

PROCEDURE

Part A. Observation

1. Place a preserved spider, crayfish, and grasshopper in a dissecting pan. **CAUTION:** *Wear protective gloves when handling preserved specimens.*

2. Examine the following features of the three arthropods. Use Figure 1 as a guide. Record your observations in Table 1.

Segmented Body

Examine both the dorsal (top) and ventral (underside) surfaces of the specimens' bodies. Do the animals appear to be segmented in any body region?

Type of Skeleton

Touch and lightly squeeze the animal. If the outside is hard, the animal has an external skeleton. If soft, the animal has an internal skeleton. Is the skeleton external or internal?

Types of Appendages

An appendage is any limblike structure that projects from a body or an organ. See if the segments that

make up the appendages will bend. If appendages bend between segments, they are jointed. Are the appendages "jointed"?

Modifications of Appendages

Are the appendages structurally and functionally different? List some of the functions of the various appendages. Use your text to help you.

Wings

Does the animal have wings?

Number of Body Regions

Examine the dorsal and ventral surfaces. Look for the head, thorax, and abdomen. How many body regions does each of the specimens have?

Fusion of Body Regions

Are the head and thorax fused together to form a cephalothorax region? Look for a groove or depression between the two. This groove suggests that they were once separate regions.

Comparing Arthropods

Figure 1

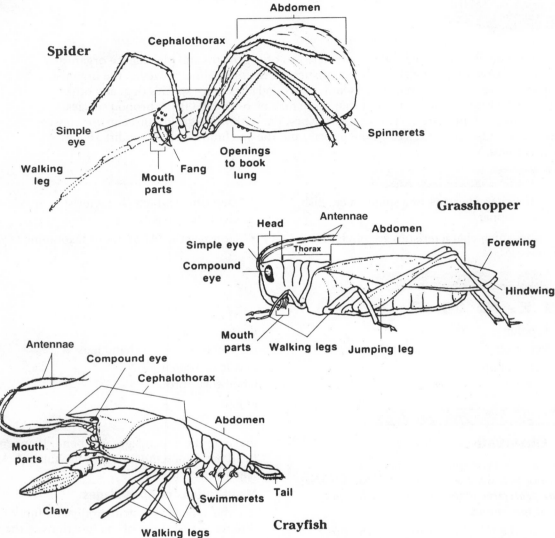

Spider

Abdomen

Cephalothorax

Simple
eye

Walking
leg

Mouth
parts

Fang

Openings
to book
lung

Spinnerets

Grasshopper

Antennae

Head

Abdomen

Simple eye

Thorax

Forewing

Compound
eye

Hindwing

Mouth
parts

Walking legs

Jumping leg

Antennae

Compound eye

Cephalothorax

Abdomen

Mouth
parts

Claw

Walking legs

Swimmerets

Tail

Crayfish

PROCEDURE continued

Number of Eyes

How many simple eyes and compound eyes does the animal have? (You will need to examine the specimens with a hand lens or stereomicroscope.)

Number of Antenna Pairs

Does the animal have antennae present? If so, how many pairs of antennae does the animal have?

Number of Leg Pairs

How many pairs of legs does each animal have? (Count the large claw structures on the crayfish

as legs. The small leglike appendages on the crayfish abdomen are not legs. They are swimmerets and function in reproduction.)

3. Wash your hands thoroughly with soap and water after handling preserved materials.

Lab
28-1

Comparing Arthropods

Part B. Making a Key

1. Prepare a taxonomic key that can be used to identify organisms that belong to these three classes. For assistance, consult with your teacher or refer to *Organizing Information* in the *Skill Handbook* of your textbook. Write your key in the space provided in Data and Observations.

DATA AND OBSERVATIONS

Table 1

Characteristics of Arthropods			
Characteristic	**Crustacea (crayfish)**	**Insecta (grasshopper)**	**Arachnida (spider)**
Segmented body?			
Type of skeleton?			
Type of appendages?			
Appendages structurally different?			
Functions of appendages?			
Wings present?			
Number of body regions?			
Body regions fused?			
Number of simple eyes?			
Number of compound eyes?			
Number of antenna pairs?			
Number of leg pairs?			

Student Key:

Comparing Arthropods

ANALYSIS

1. Which of the characteristics are the same in all three animals?

_____

_____

2. In which characteristics do the three animals appear to differ?

_____

_____

3. At which level in the classification hierarchy (phylum or class) do members of a group have more characteristics in common?

_____

_____

4. The classification systems used today attempt to reflect genetic relationships. Why can you say that members of a group are more closely related to one another than they are to members of another group?

_____

_____

_____

_____

FURTHER EXPLORATIONS

1. Examine centipedes or millipedes, or photographs of these arthropods. What characteristics are shared with the other classes of arthropods? What characteristics are different?

2. Investigate how to build a spider-web box. Build a spider-web box and observe web-building behavior.

INVESTIGATION

How Does Temperature Affect Mealworm Metamorphosis?

Moths, butterflies, bees, and beetles are examples of insects that have four stages in their life cycles: egg, larva, pupa, and adult. This sequence of events is called complete metamorphosis. The female adult lays eggs, which hatch into larvae. Larvae eat an enormous amount of food and grow very rapidly. When a larva reaches a certain size, it enters an immobile stage called a pupa. Inside its protective covering, the pupa undergoes many changes controlled by hormones. When metamorphosis is complete, an adult emerges from the pupa. The mealworm, *Tenebrio*, is an excellent insect for the study of complete metamorphosis.

OBJECTIVES

- Observe the four stages of the life cycle of the mealworm, *Tenebrio*.

- Hypothesize how an increase in temperature will affect the mealworm's rate of metamorphosis.

- Conduct an experiment to test the effect of temperature on the emergence of an adult mealworm from a pupa.

MATERIALS

samples of mealworms (egg, larva, pupa, adult)	mealworm pupae (of same age) (4) wax marking pencil	stereomicroscope plastic vials (4) foam plugs (4)	incubator (at 30°C) thermometer laboratory apron goggles

PROCEDURE

1. Examine samples of the four stages of mealworms under the stereomicroscope. Relate them to the life cycle of *Tenebrio* shown in Figure 1.

2. Make a hypothesis that predicts how an increase in temperature will affect the length of time it takes a mealworm pupa to become an adult. Write your hypothesis in the space provided.

3. With your marking pencil, label the four plastic vials "Room Temp. A," "Room Temp. B," "30°C A," and "30°C B." These labels indicate the temperature at which the pupae will be stored. Label them also with your name (or group name) and the date.

4. Place one pupa in each of the four vials and stopper with foam plugs. The foam plugs will allow the insects to breathe.

5. Store your vials at their proper temperatures with those of the rest of the class. Record the starting date in Table 1. Record the actual room temperature in Table 2.

6. Check your vials daily for the presence of adult mealworms. When you observe an adult in a vial,

Figure 1

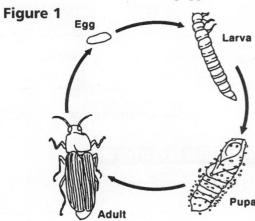

record in Table 1 the number of days needed for emergence.

7. Dispose of the adults after they emerge and clean your vials following the instructions of your teacher.

8. Wait for metamorphosis to be complete for mealworms of the entire class, compile the class data, and complete Table 2. Calculate the average time for emergence by dividing the total number of days by the total number of pupae.

How Does Temperature Affect Mealworm Metamorphosis?

HYPOTHESIS

_____

_____

DATA AND OBSERVATIONS

Table 1

Tenebrio Metamorphosis		
Temperature	Starting date	Length of time for emergence (days)
Room temp. A		
Room temp. B		
30°C A		
30°C B		

Table 2

Calculations			
Temperature	Total number of days for entire class	Total number of pupae for entire class	Average time for emergence
Room temp. (_____°C)			
30°C			

ANALYSIS

1. How did an increase in temperature affect the time needed for emergence? _____

2. Why do you think a temperature increase causes this effect? _____

_____

3. Why might the class averages be a more accurate measurement of the time needed for emergence than

your data alone? _____

4. Sequence the stages of the *Tenebrio* life cycle. _____

CHECKING YOUR HYPOTHESIS

Was your **hypothesis** supported by your data? Why or why not?

_____

_____

FURTHER INVESTIGATIONS

1. Repeat the Investigation using other insects, such as *Drosophila*, to see if temperature affects their metamorphosis.

2. Design an experiment to test the effect of temperature on other stages in the life cycle of *Tenebrio*.

Lab
29-1

Sea Star Dissection

Phylum Echinodermata is a group of invertebrate marine animals that includes sea stars, sea urchins, and sand dollars. "Echinos" means spiny, and "derma" means skin. Sea stars are the most common representative animals of this phylum. Both internally and externally, the anatomy of sea stars in the genus *Asterias* is characterized by five arms and a five-part anatomical arrangement.

OBJECTIVES

- Dissect a preserved sea star.
- Identify and label the major external and internal structures of a sea star.

- Describe how the water vascular system enables a sea star to move.

MATERIALS

scissors
preserved sea star
dissecting tray
dissecting probe

stereomicroscope
slide of gills and pedicellariae
plastic or rubber gloves
laboratory apron
goggles

PROCEDURE

Part A. External Anatomy

1. Place your sea star in the dissecting tray so that the top surface faces upward as shown in Figure 1. **CAUTION:** *Wear protective gloves when handling preserved specimens.*

2. Examine the animal's top surface. Locate the **central disc** and the five arms, or **rays**, that extend from the central disc, as shown in Figure 1.

3. Locate the **madreporite plate**. It is a round, sievelike structure that looks almost like a wart. Use Figure 1 as a reference to find the madreporite plate.

4. Note the many **spines** scattered over the surface of the arms and the central disc. These spines are attached to the plates of the sea star skeleton just under the skin. These plates are called ossicles.

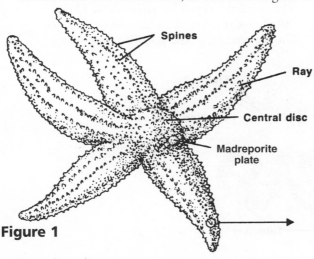

Figure 1

Figure 2

Copyright © Glencoe/McGraw-Hill, a division of The McGraw-Hill Companies, Inc.

Sea Star Dissection

PROCEDURE continued

5. Examine the demonstration slide that your teacher has set up on the stereomicroscope. Among the spines are small pincerlike pedicellariae and gills, as shown in Figure 2. The **pedicellariae** keep the surface of the skin free of debris, while the **gills** are involved in respiration.

6. Turn the sea star over. Use Figure 3 to help you locate the **mouth** and the five **ambulacral grooves** that extend from the mouth along the middle of each ray. Numerous **tube feet** used for locomotion and absorption of dissolved oxygen are present along the grooves.

Part B. Internal Anatomy

Digestive System

1. Once again, place the sea star in the dissecting tray so that the top surface faces upward.

2. Carefully cut a ring around the madreporite plate.

3. Use your scissors to cut off the tip of any arm except for the two arms next to the madreporite plate.

4. Starting at the end of the arm with its tip cut off, use your scissors to remove the remaining skin/skeleton from the top of the arm and from the central disc. This will expose the sea star's internal organs. Start your cut at the end of the cut arm.

5. Use Figure 4 to locate the **digestive gland,** a large olive-green gland with two branches that fills most of the arm .

6. Draw and label this digestive gland on the outline of the sea star in Figure 6.

7. Turn your sea star over and locate the **mouth**. The mouth is attached to the pouchlike **stomach**, which can be seen through the opening you have cut in the top surface.

Figure 3

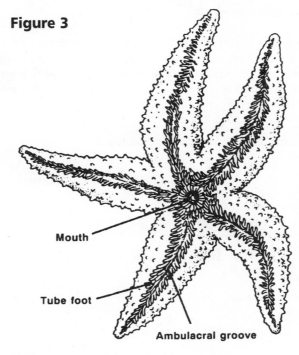

Mouth

Tube foot

Ambulacral groove

8. Draw and label the mouth and stomach on the sea star in Figure 6.

9. Locate the short tube that connects the stomach to the digestive glands.

Reproductive Organs

1. Remove the entire digestive gland from the dissected arm of your sea star.

2. Locate the pale, lumpy organs under the digestive gland near the central disc as shown in Figure 4. These are the reproductive organs. They are called **gonads.** Sea stars have separate sexes. During spawning, the gonads are very large,

Figure 4

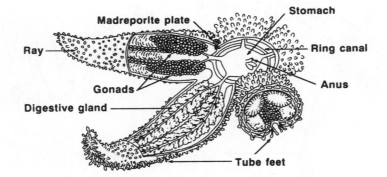

Madreporite plate

Ray

Gonads

Digestive gland

Stomach

Ring canal

Anus

Tube feet

Sea Star Dissection

but in preserved specimens, they are usually quite small. The male and female gonads look very much alike in preserved specimens. In living sea stars, the testes are gray and the ovaries are orange.

3. Draw and label the gonads in Figure 6.

Water Vascular System

1. Carefully remove the reproductive organs and the remaining parts of the digestive system (including the stomach). This will expose the **water vascular system**. Be careful not to damage the madreporite plate.

2. Study your sea star. Use Figure 5 to find each of the structures of the water vascular system that are in bold print in the following paragraph.

Water enters the system through the **madreporite plate**. The madreporite plate is connected to the circular **ring canal** by the **stone canal**. The water is then distributed to the **radial canals** that are in each of the rays. These canals deliver water to the **tube feet**. The tube feet contain **ampullae** (bulblike structures). By alternating between contracting and expanding the ampullae, the sea star is able to move. As the ampullae contract, they force water into the tube feet, and the tube feet lengthen. The sea star places the lengthened tube feet in the direction it is

going. Then the ampullae relax and expand. When this happens, water leaves the tube feet, thus shortening each tube foot and creating suction at its end.

3. Trace the pathway that seawater takes from the madreporite plate to the tube feet. Mark this pathway on Figure 5 with short dashed lines.

4. After completing the dissection, dispose of your sea star according to your teacher's instructions. Wash your hands thoroughly with soap and water.

Figure 5

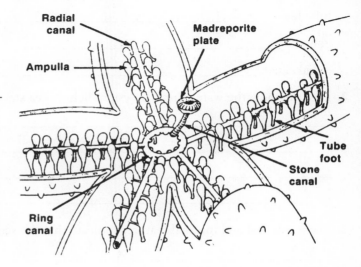

DATA AND OBSERVATIONS

Figure 6

Sea Star Dissection

ANALYSIS

1. List the functions of the:

madreporite plate. _____

spines. _____

gills. _____

digestive glands. _____

pedicellariae. _____

tube feet. _____

gonads. _____

radial canal. _____

2. Describe how the water vascular system helps a sea star move around.

_____

_____

_____

_____

3. What external and internal features indicate that the sea star is organized in a pattern of fives?

_____

_____

_____

FURTHER EXPLORATIONS

1. Demonstrate how a plastic dropper can be used as a model to show the action of tube feet and ampullae as the dropper is filled and emptied of water.

2. Examine the anatomies of other echinoderms (sea cucumbers, crinoids, sea urchins, sand dollars) and note their similarities and differences with the sea star anatomy.

INVESTIGATION

How Do Sea Stars Respond to Gravity?

Investigators have shown that sea stars move by pushing with their tube feet as well as by pulling with them. A sea star moves, or walks, in response to external stimuli, such as the force of gravity. Geotaxis, the response to the force of gravity, can be either positive (in the same direction as the force of gravity) or negative (in the opposite direction as the force of gravity).

OBJECTIVES

- Observe the external structures of a living sea star.
- Observe the walking behavior of a living sea star.
- Make hypotheses that describe a sea star's response to gravity.
- Determine how a sea star responds to the force of gravity.

MATERIALS

living sea star
seawater

colored celluloid
glass container for
  sea star

glass plate
drawing paper

laboratory apron
goggles

PROCEDURE

1. Carefully place a living sea star, top surface up, in a glass container of seawater.

2. Make drawings on a sheet of drawing paper of the top and bottom surfaces of the sea star. Refer to the figures in Exploration 29-1. Correctly draw and label the following.

   top surface—central disc (central body)
             rays (five arms)
             madreporite plate (round structure)
   bottom surface—mouth
             ambulacral grooves (five)
             tube feet

The tube feet are connected internally to a canal system that in turn connects to the madreporite plate. Water filters into the canal system through the madreporite plate. Within the canals, water is under pressure and helps to extend and retract the tube feet. The extending and retracting of the tube feet is used by the sea star for locomotion.

3. Gently remove the sea star from the container. Place the piece of celluloid in the water at the bottom of the container. Place the sea star on top of the piece of celluloid. See Figure 1.

4. Observe the stepping action of the sea star on the celluloid sheet by looking at the underside of the container while another student holds it up.

Observe the movement of the celluloid sheet. After completing your observations, carefully remove the celluloid from the sea star. Write a description of the movement in Data and Observations.

5. Make **hypotheses** that describe how a sea star will react to (1) being turned over and (2) being placed on an inclined surface. Operate under the assumption that the reactions will be responses to the force of gravity. Write your hypotheses in the space provided.

6. Turn the sea star over in its container so that the bottom surface faces up. Note the reaction of the sea star. Record your observations in Data and Observations.

Figure 1

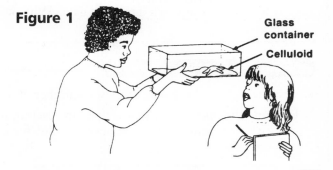

Glass
container
Celluloid

How Do Sea Stars Respond to Gravity?

PROCEDURE continued

7. Lean one edge of the glass plate on the bottom of the container. Allow the sea star to crawl onto the plate. When it is well onto the plate, tilt the plate at a 45 degree angle. See Figure 2.

8. Observe how the sea star reacts to being on an inclined surface. Record your observations.

9. After several minutes, shift the glass plate so that the edge that was once down is now up, as in Figure 2. Observe the reaction of the sea star. Record your observations in Data and Observations.

Figure 2

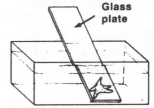

Glass plate

10. After making your observations, return the sea star to its aquarium. Wash your hands thoroughly with soap and water after handling the sea star.

HYPOTHESES

_____

_____

DATA AND OBSERVATIONS

1. Describe what happened to the celluloid as the sea star moved. _____

_____

2. Describe the movements of the sea star after it was turned over. _____

_____

3. How many rays were attached to the dish before the body of the animal turned over? _____

4. How did the sea star react to finding itself on an inclined surface? _____

5. How did the sea star react to the reversal of "up" and "down"? _____

ANALYSIS

1. Explain how the sea star moved over the celluloid. _____

2. Does a sea star show positive or negative geotactic behavior? _____

CHECKING YOUR HYPOTHESES

Was your **hypothesis** supported by your data? Why or why not? _____

_____

FURTHER INVESTIGATIONS

1. Test the righting response of a sea cucumber or a sea urchin.

2. Examine the movements of a sea star in reaction to the presence of a clam. Sea stars that are hungry will eat a live clam. It may take several hours for a sea star to get the clam open and extend its stomach into the clam.

EXPLORATION   Examining Fish Scales

There are four types of fish scales. Placoid scales are heavy, toothlike scales on cartilaginous fishes. Ganoid scales are thick and covered by an enamel-like substance. Most fish, including trout, perch, and salmon have either cycloid or ctenoid (TEA noid) scales. The approximate age of a fish with ganoid, cycloid, or ctenoid scales can be determined by examining the pattern of lines or rings, called circuli, on its scales. The number of dark rings is an estimate of the fish's age.

OBJECTIVES

• Observe different types of fish scales under the microscope.

• Determine the type of fish scale.

• Estimate the age of different fishes.

MATERIALS

scales from five different fish   dropper
microscope slides   water
cover slips   forceps
compound light microscope   paper towels

PROCEDURE

Part A. Type of Scale

1. Obtain a fish scale from your teacher. Using the forceps, examine the scale and compare it to the scales shown in Figure 1.

2. Describe the size, shape, thickness, flexibility, and any distinguishing features of the scale in the Table 1.

3. Place the scale in the center of a slide. Use the dropper to place a drop of water on the scale. Cover the scale with a cover slip.

4. Place a paper towel at the edge of the coverslip to remove excess water. **CAUTION:** *Use caution when handling microscope slides and cover slips.*

5. Place the slide you prepared under the microscope and focus it under low power.

6. Use Figure 1 and Figure 30.4 on page 797 of your textbook to determine whether the scale is placoid, ganoid, cycloid, or ctenoid.

7. Record your data in Table 1.

Figure 1

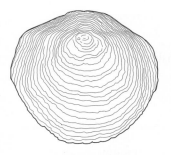

Cycloid Scale

Ganoid Scale

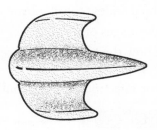

Placoid Scale

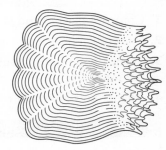

Ctenoid Scale

Examining Fish Scales

Figure 2

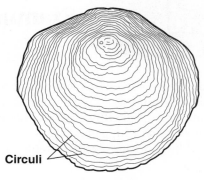

Circuli

Cycloid Scale

PROCEDURE continued

Part B. Estimate Fish Age

8. With the microscope still on low power, locate the circuli on the fish scale. Each dark line represents one year of the fish's life.

9. Count the lines to estimate the age of the fish. Use Figure 2 as a reference.

10. Record your data in Table 1.

11. Dispose of the fish scale as directed by your teacher.

12. Repeat steps 1–11 with the other four fish scales.

13. Wash your hands thoroughly with soap and water after handling biological materials.

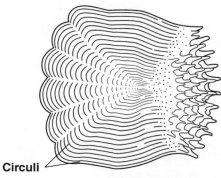

Circuli

Ctenoid Scale

DATA AND OBSERVATIONS

Table 1

Scale Sample	Description of Scale	Type of Scale	Estimated Age of Fish (years)
1			
2			
3			
4			
5			

ANALYSIS

1. Compare placoid and ganoid scales to cycloid and ctenoid scales.

2. Infer possible adaptive advantages of the different fish scale types.

3. Form a hypothesis about how the type and arrangement of scales could affect a fish's movement through the water.

4. Predict how an injury to or infection of scales could affect a fish.

FURTHER EXPLORATIONS

1. Investigate how the type and arrangement of scales are related to a fish's niche and habitat.

2. Research how fish repair or replace injured scales.

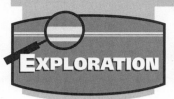

EXPLORATION Frog Dissection

Amphibians are more complex than fishes. The amphibians' lungs and four legs helped to make life on land possible. By examining the anatomy of amphibians, one can see many structures that are basic to both amphibians and all the more advanced vertebrates.

OBJECTIVES

- Identify the external organs of a preserved frog.
- Dissect a frog to identify the frog's internal organs.
- Explain the adaptive advantages of some of the frog's structures.

MATERIALS

frog (preserved)
scissors
dissecting pins (6–10)
dissecting pan

stereomicroscope
dissecting probe
forceps
microscope slides (2)

coverslip
water
metric ruler
compound light
  microscope

plastic or rubber gloves
laboratory apron
goggles

PROCEDURE

Part A. External Anatomy

Refer to Figure 1 for this part of the Procedure.

1. Put on a laboratory apron. Rinse a preserved frog well with water. Place it ventral (bottom) surface down in your dissecting pan. **CAUTION:** *Wear protective gloves when handling preserved specimens.*

2. Note the arrangement of the spots and the coloration of the frog. The color of the frog is caused by scattered granules in the epidermis and chromatophore cells in the dermis. Chromatophore cells are cells that contain pigments.

3. Remove a 1 cm × 1 cm section of the dorsal (top) skin containing one of the frog's spots. Make a wet mount of this piece of skin.

4. Place the wet mount on the compound microscope under low power and then high power. Chromatophores are usually star-shaped. Dispersal of the pigment into the rays makes the skin darker. When the pigment is concentrated in the center of the chromatophores, the skin is lighter.

5. Make a drawing of the chromatophores in the space provided in Data and Observations.

Figure 1

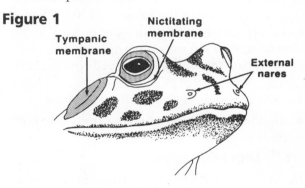

6. Locate the thin membrane that covers the eye from below. This is the **nictitating membrane**, which protects the eye when the frog is under water and keeps it moist when the frog is on land.

7. Notice the large **tympanic membranes** behind the eyes. These membranes function as eardrums to receive sound waves. The **external nares** are the frog's nostrils.

8. Examine the forelegs and hindlegs of the frog, noting the number of toes. If your frog is a male, it will have roughened pads near the thumbs. These are used to hold the female during mating.

9. Answer questions 1, 2, and 3 in the Analysis.

10. Find in Figure 1 the structures that are in bold print in Part A.

Frog Dissection

Part B. Oral Cavity

Refer to Figure 2 for this part of the Procedure.

1. Open the mouth by cutting the jaws with the scissors. Locate the **maxillary teeth** around the edge of the jaw. They hold food but are not used for chewing.

2. Locate the slitlike **glottis** at the back of the throat. This opening leads to the respiratory system. Above the glottis is the **opening to the esophagus**.

3. Find the **eustachian tubes** at the posterior corners of the upper jaw. Probe with your dissecting probe to find out where they lead. These tubes equalize pressure within the ear.

4. Locate in a male frog the opening that leads to the **vocal sacs** at the widest corner of the lower jaw. They amplify the male's mating call.

5. Notice the shape of the **tongue** and where it is attached. It can be flipped forward to catch prey.

6. Locate the **nostril openings** in the roof of the mouth. Between the nostril openings are two **vomerine teeth**. Feel these teeth and the maxillary teeth with your fingers.

7. Answer questions 4 and 5 in the Analysis.

8. Find in Figure 2 the structures that are in bold print in Part B.

Part C. Digestive System

Refer to figures 3 and 4 for this part of the Procedure.

1. Place your frog dorsal surface down in the dissecting pan. Open the frog by cutting the skin around the abdomen in the manner shown in Figure 3.

2. Insert the point of your scissors just through the muscle above the anal opening and make a cut extending to the lower jaw. Cut sideways at both the forelegs and hindlegs as shown in Figure 3. Pin the muscles down to the dissecting pan. **NOTE:** If your frog is female and contains black eggs, they must be removed carefully before the internal organs can be observed.

3. Locate the **esophagus** as shown in Figure 4. Pass a probe into the **stomach**. Note that the lower end of the stomach is constricted. This

Figure 2

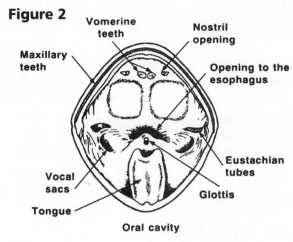

Vomerine teeth · Nostril opening · Maxillary teeth · Opening to the esophagus · Eustachian tubes · Glottis · Oral cavity · Tongue · Vocal sacs

Figure 3

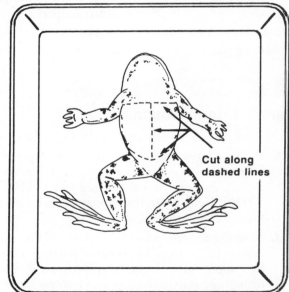

Cut along dashed lines

constriction is the **pyloric sphincter**. It regulates the amount of food that enters the small intestine.

4. Cut open the stomach and observe its lining. If food is present in the stomach, try to identify it. Usually insect body parts will be present.

5. Follow the digestive tract beyond the pyloric sphincter to the coiled **small intestine**. The first portion that usually runs parallel to the stomach is the **duodenum**.

6. Cut open the lower part of the small intestine. Place a piece of the small intestine on a clean glass slide with the inside surface up. Observe

_____

Frog Dissection

Lab
30-2

the surface of the small intestine under a stereomicroscope. Note the villi, the many small folds in the lining that increase the absorption of nutrients into the bloodstream.

7. Follow the digestive tract below the small intestine where it widens into the large intestine, or **colon**. The colon ends in the **rectum**, which in turn opens into the **cloaca**. The cloaca opens to the outside of the frog. The digestive, reproductive, and excretory systems all open into the cloaca.

8. Note the large brown **liver**. Lift the lobes of the liver to locate the green **gallbladder**. The gallbladder stores bile that is secreted by the liver. Bile aids in the digestion of fats.

9. Locate the **pancreas**, a soft, irregular, pinkish organ that produces digestive enzymes, found lying in a membrane between the stomach and duodenum.

10. Answer questions 6 and 7 in the Analysis.

11. Find in Figure 4 the structures that are in bold print in Part C.

Part D. Respiratory and Circulatory Systems

Refer to Figure 4 for this part of the Procedure.

Air is drawn into the mouth by expansion of the throat. The external nares close, then the throat muscles contract and air is forced into the lungs through the glottis. Air is expelled as the nares remain closed, the throat expands, and air enters the mouth again from the lungs. The glottis closes, the nares open, and the throat contracts, forcing the air out through the nares.

1. Probe the glottis to see where it leads. Locate the **trachea**, the passageway between the glottis and the lungs.

2. Locate the pinkish-gray **lungs**.

3. Notice the three-chambered **heart** between the lungs and posterior to the trachea. The pointed **ventricle** is lighter in color than the two thin-walled **atria**.

4. Lift the stomach and find the **spleen**, a round red organ. The spleen filters the blood, taking out improperly functioning red blood cells.

5. Answer questions 8 and 9 in the Analysis.

6. Find in Figure 4 the structures that are in bold print in Part D.

Part E. Excretory and Reproductive Systems

Refer to Figure 4 for this part of the Procedure.

1. Examine the **kidneys** that lie against the dorsal body wall in the posterior region of the body cavity. Each kidney has a yellow stripe, known as the **adrenal body**, that secretes hormones. The kidneys filter the blood and produce urine, which drains into the **urinary bladder**, a thin-walled bag that attaches to the cloaca.

2. A female frog has two lobed, grayish ovaries that lie close to the kidneys. In a mature female, the two ovaries might be filled with black and white eggs.

3. Locate in a male frog the white **testes** that can be found close to the kidneys. Look at the reproductive organs of both sexes.

4. Examine the yellow, fingerlike **fat bodies** attached near the kidneys. Compare their size with those in a frog of the opposite sex. The fat bodies provide nourishment for the gametes.

5. Answer question 10 in the Analysis.

6. Find in Figure 4 the structures that are in bold print in Part E.

Part F. The Brain

Refer to figures 5 and 6 for this part of the Procedure.

1. Turn over your frog so that the dorsal side once again faces up.

2. Insert the point of your scissors through the skin at the base of the head and remove the skin from the head area.

3. Bend the frog to determine the approximate region of the "neck."

4. Insert your scissors and clip across the upper spinal cord in the region of the neck.

5. Locate the white **spinal cord** enclosed within the vertebrae.

6. Use your forceps to remove the bone above the spinal cord, working forward until you have reached the nostril area. You will be exposing the brain, as shown in Figure 6.

7. Locate the spinal cord and the **olfactory lobes**, **cerebrum**, **optic lobe**, **cerebellum**, and **medulla oblongata** of the brain, using Figure 6 as a guide.

8. Answer question 11 in the Analysis.

Frog Dissection

Figure 4

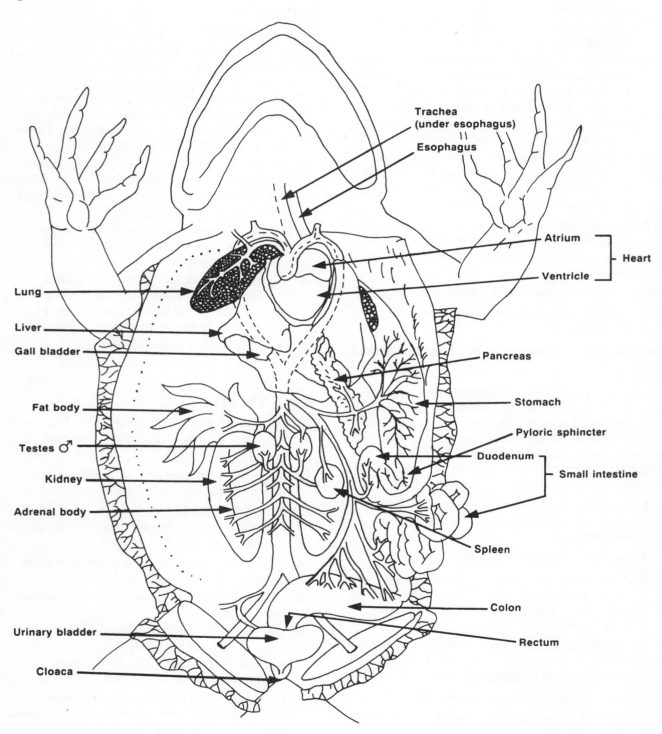

Trachea
(under esophagus)

Esophagus

Atrium

Ventricle

Heart

Lung

Liver

Gall bladder

Pancreas

Stomach

Fat body

Pyloric sphincter

Testes ♂

Duodenum

Small intestine

Kidney

Adrenal body

Spleen

Colon

Urinary bladder

Rectum

Cloaca

Frog Dissection

Figure 5

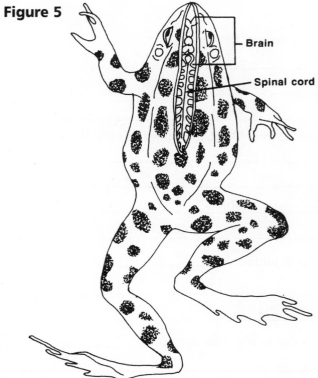

— Brain

— Spinal cord

Figure 6

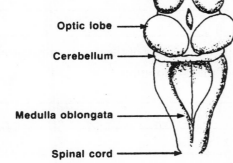

Olfactory lobe

Cerebrum

Optic lobe

Cerebellum

Medulla oblongata

Spinal cord

DATA AND OBSERVATIONS

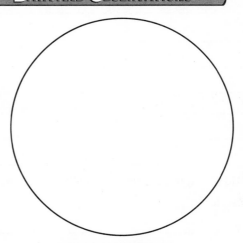

Chromatophores

ANALYSIS

1. a. Compare the position of the frog's eyes with the position of human eyes.

ANALYSIS continued

b. How is the positioning of the frog's eyes an adaptive advantage for the frog?

_____

_____

_____

2. Compare the colors of the dorsal and ventral surfaces of the frog. Of what adaptive value to the frog is each of these colorations? _____

_____

_____

_____

3. Is the dispersal of the pigment in all of the chromatophores uniform? _____

4. How is the tongue attached? _____

5. To what structures do the eustachian tubes lead? _____

6. How does the lining of the stomach compare with the lining of the small intestine? _____

_____

7. How many lobes does the frog's liver have? _____

8. Where does the glottis lead? _____

9. Sequence the passage of air into and out of a frog. _____

_____

_____

10. Are the fat bodies larger in male or female frogs? _____ Why is this so?

_____

_____

11. The largest parts of the frog's brain are the olfactory lobes and optic lobes, the centers of smell and vision, respectively. How is this adaptation an advantage for the frog's lifestyle? _____

_____

_____

FURTHER EXPLORATIONS

1. Find out what kinds of amphibians can be found in the area in which you live.

2. Observe the behavior of a living frog in an aquarium. Observe how it floats, swims, and uses its eyes, as well as other behaviors.

Lab
31-1

INVESTIGATION **Modeling Reptile Skin Shedding**

Reptile skin is thick, dry, and scaly, and provides protection from dehydration, predators, and injury. As reptiles grow, they periodically undergo ecdysis (EK de sis), the process of shedding old skin. New skin is produced beneath the old one before shedding occurs. The skins of some reptiles, such as lizards and turtles, are shed in pieces. For most snakes, the skin is shed as one piece.

OBJECTIVES

- Observe the shed skins of different reptiles under the microscope.
- Model reptile shedding.

MATERIALS

shed skins of several reptiles   latex gloves (4)
microscope slides   cornstarch
coverslips   vegetable oil
compound light microscope   bowls, 1-quart (2)
dropper   stopwatch or clock with a second hand
water   rinse water
forceps   goggles
paper towels   laboratory apron

PROCEDURE

Part A. Examining Shed Reptile Skins

1. Using the scissors, cut a section from a shed reptile skin. Examine the shed skin and describe the thickness, flexibility, and any distinguishing patterns or features in the Table 1.

2. Use the forceps to place the section of shed skin in the center of a slide. Use a dropper to place a drop of water on the shed. Cover the shed with a coverslip.

3. Place a paper towel at the edge of the coverslip to remove excess water. **WARNING:** *Use caution when handling microscope slides and coverslips.*

4. Place the slide under the microscope and focus it under low power.

5. Examine the shed skin and record your observations in Table 1.

6. Dispose of the shed skin as directed by your teacher.

7. Repeat steps 1–6 for any other sheds provided by your teacher.

8. Wash your hands thoroughly after handling biological materials.

Copyright © Glencoe/McGraw-Hill, a division of The McGraw-Hill Companies, Inc.

Modeling Reptile Skin Shedding

PROCEDURE continued

Part B. Modeling Reptile Ecdysis

1. Obtain two bowls and add cornstarch to one and vegetable oil to the other as directed by your teacher.

2. Obtain four latex gloves.

3. Place a glove on your hand.

4. Think of what a snake might do to remove its old skin from its body. Try to remove the glove from your hand without someone else's help or without using your other hand, arms, or the rest of your body. Have your partner time how long it takes you to remove the glove. Record the time, in seconds, in Table 2.

5. Wash and dry your hands after removing the glove.

6. Using the same hand from step 3, use the wash bottle to wet your hand up to your wrist.

7. Without drying your hand, put on a new glove.

8. Repeat steps 4 and 5.

9. Repeat steps 6–8 covering your hand to your wrist with cornstarch.

10. Repeat steps 6–8 covering your hand to your wrist with vegetable oil.

11. Switch roles with your partner and begin at step 3.

12. Clean your work area upon completion of the experiment and dispose of materials as directed by your teacher. Wash your hands thoroughly.

DATA AND OBSERVATIONS

Table 1

Sample of Shed Skin	Description
1	
2	
3	
4	
5	

Table 2

Experimental Conditon	Time (seconds)
Glove	
Glove and Water	
Glove and Cornstarch	
Glove and Vegetable Oil	

Modeling Reptile Skin Shedding

ANALYSIS

1. Under which condition was the glove easiest to remove?

_____

_____

_____

2. Snakes produce an oily substance between the surface of their new skin and the shed.
What is the purpose of this substance?

_____

_____

_____

3. What was the purpose of removing the glove without any substances on your hand?

_____

_____

_____

4. Compare your data to that of other students in your class. What could account for
differences in data?

_____

_____

_____

5. Infer why reptiles shed their skin.

_____

_____

_____

FURTHER INVESTIGATIONS

1. Investigate what might happen if a reptile could not shed successfully.

2. Research how factors such as temperature, humidity, amount of food, and injury can affect the
frequency with which a reptile sheds its skin.

EXPLORATION   Examining Bird Feet

You may have seen a sandpiper hurrying along the shore away from an oncoming wave or a sparrow perching on a telephone wire. Birds' feet are adapted for a variety of functions. Birds run, hop, walk, scratch, perch, attack, defend, take off for flight, preen, swim, or obtain food with their feet. Birds' feet have claws and are composed primarily of bone, tendon, and tough, scaly skin. Most birds have four toes arranged in a variety of ways on the foot. The toes have various sizes and shapes and are adapted to particular lifestyles. By examining a bird's foot, you may be able to tell something about where the bird lives, how it gets food, what it eats, or how it defends itself.

OBJECTIVES

- Identify adaptations of birds' feet that make them suited for particular habitats and lifestyles.

- Compare and contrast the feet of various species of birds.

- Relate the characteristics of birds' feet to their functions and to the birds' habitat.

MATERIALS

field guide to birds

PROCEDURE

1. Examine Figure 1 and note the various parts of a bird's foot.

2. Examine the diagrams of birds' feet in Table 1. Fill in the "Structural adaptations" column. Describe the features of each foot type that make it suited for its particular function. Look for the following traits.

   a. number of toes

   b. length of toes

   c. position of toes: all facing forward or three in front; curled or straight

   d. width of toes

   e. presence of webbing between toes

   f. length and thickness of claws

   g. shape of claws

   h. covering on foot consisting of scales, feathers, and so on

3. Look at the drawings of different birds in Table 2. Compare their feet to those pictured in Table 1.

Figure 1

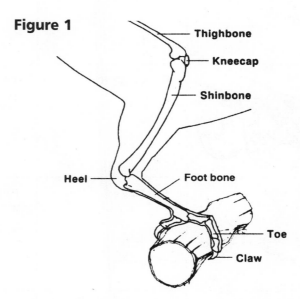

4. Write in the appropriate column of Table 2 the function of these birds' feet. Also write in Table 2 what habitat you think these birds might live in. Use Table 1 and a field guide to birds to help you.

Copyright © Glencoe/McGraw-Hill, a division of The McGraw-Hill Companies, Inc.

Examining Bird Feet

DATA AND OBSERVATIONS

Table 1

Adaptations of Birds' Feet			
Type of feet	Function	Habitat	Structural adaptations
A.	Walking, scratching, maintaining body heat	On the ground in prairies	
B.	Running	Hard, flat ground	
C.	Swimming	Lakes, ponds, and streams	
D.	Grasping prey	Meadows with streams	
E.	Perching	Meadow and woodlands	
F.	Climbing	Bark of trees	

Examining Bird Feet

Table 1 (continued)

Adaptations of Birds' Feet			
Type of feet	**Function**	**Habitat**	**Structural adaptations**
G.	Wading	Ponds with soft, muddy bottoms	
H.	Clinging	Vertical surfaces of cliffs	
I.	Full-time flight	In the air	
J.	Swimming, pushing (for a tobagganing-type sliding move-ment on breast and belly)	Water, ice, and snow	
K.	Walking on water plants	Shorelines of lakes	
L.	Defense	Various habitats	

Name Date Class

Examining Bird Feet

Lab 31-2

Data and Observations

Table 2

Birds and Their Feet			
Bird	**Close-up view of foot**	**Feet functions**	**Habitat**
Penguin 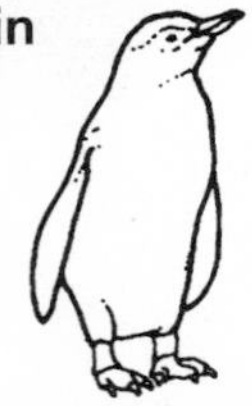	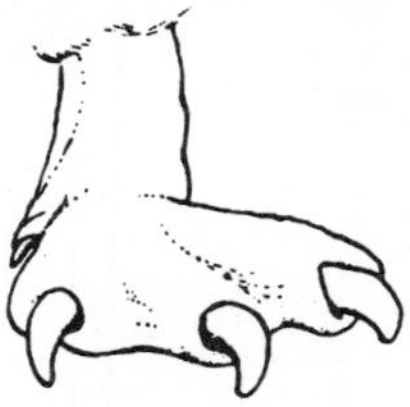		
Woodpecker 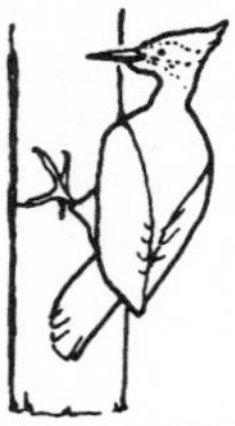	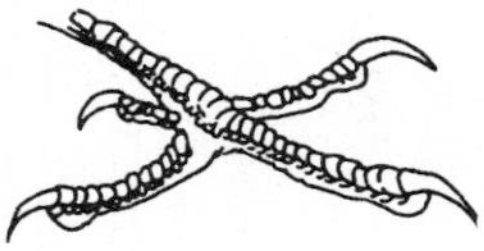		
Osprey 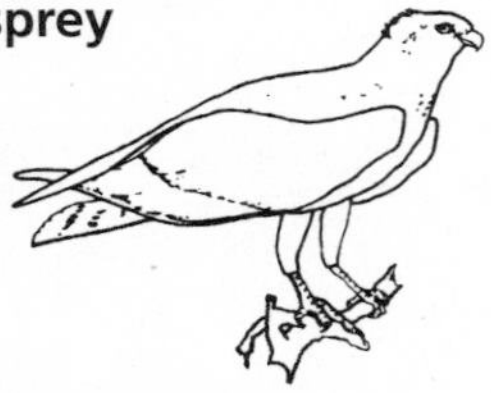	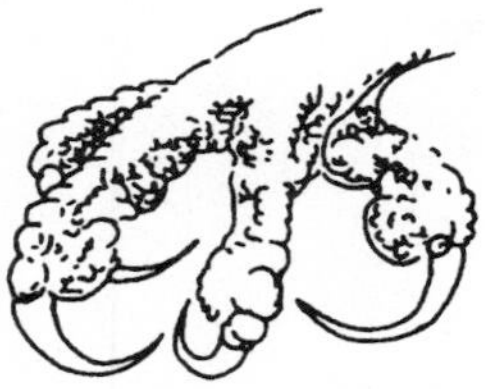		
Heron 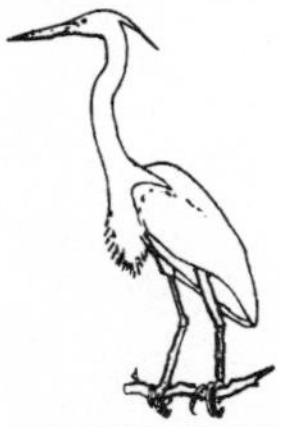	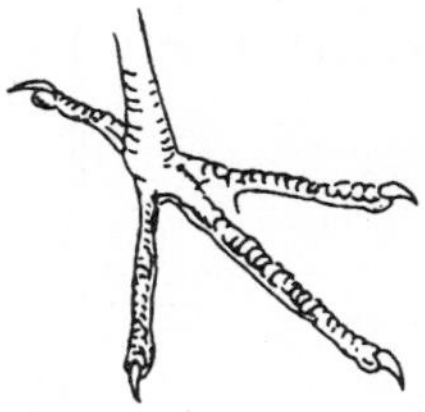		
Ruffed grouse 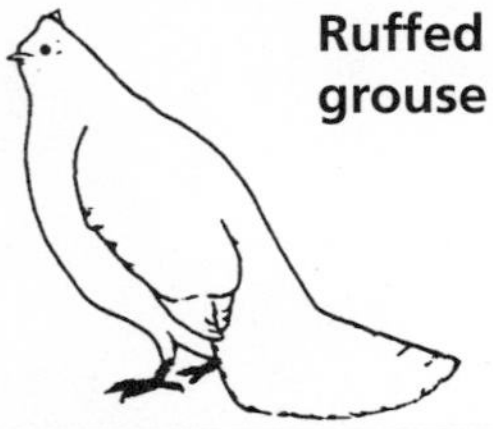	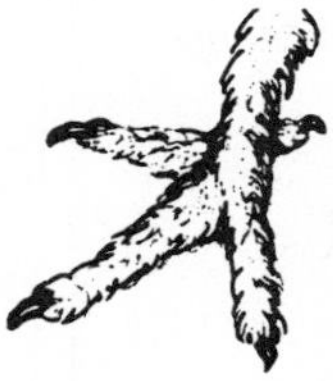		
Ostrich 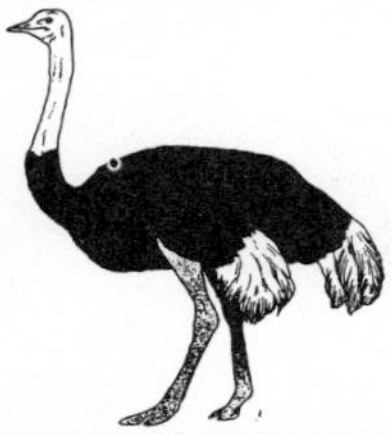	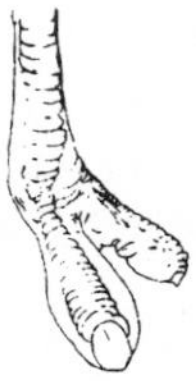		

Lab

Examining Bird Feet

31-2

Table 2 (continued)

Birds and Their Feet			
Bird	**Close-up view of foot**	**Feet functions**	**Habitat**
Hummingbird 	4× actual size 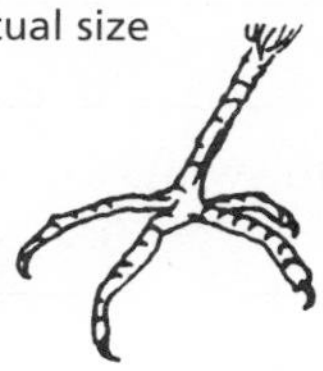		
Cliff swallow 	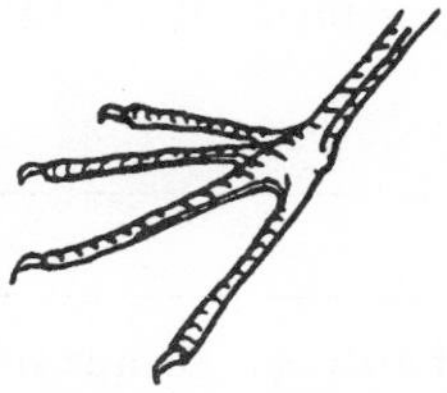		
Jacana 	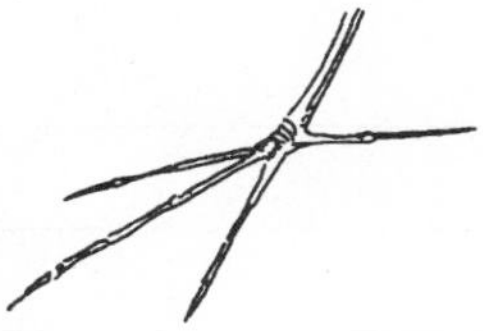		
Robin 	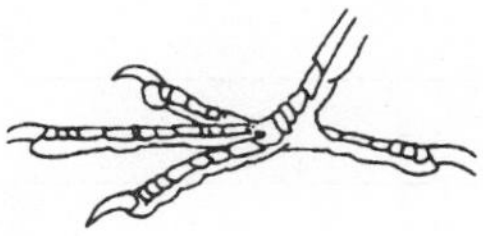		
 Mallard duck	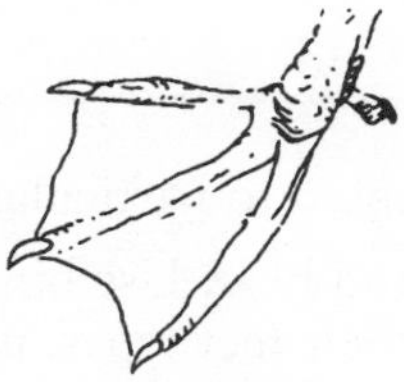		
Pheasant 	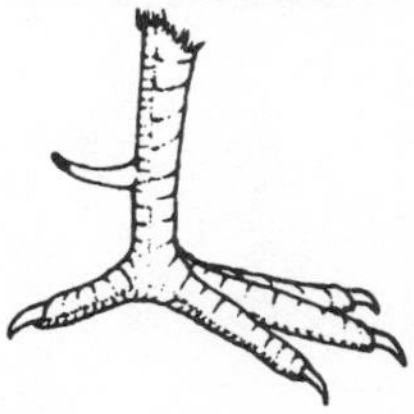		

Examining Bird Feet

ANALYSIS

1. Birds that run on the ground have toes that all point forward. Why might this be an advantage to a running bird?

_____

_____

_____

2. How does webbing on the foot of a swimming bird help it to swim?

_____

_____

3. What is the adaptive advantage of having very long toes for birds that walk on water plants?

_____

_____

4. In the wild, birds perch on branches of varying size and hardness. As a result, their claws wear naturally as they grow and their feet are well-exercised. What might happen to a pet bird that sits on one plastic perch in its cage all year?

_____

_____

_____

5. Many zoos are using rubberized netting or artificial carpets in aviaries instead of wire mesh on the floor. How might this be beneficial to the birds?

_____

_____

_____

FURTHER EXPLORATIONS

1. Look at pictures of birds' beaks to see how they are adapted to particular lifestyles.

2. Go on a bird-watching trip. Identify the birds you see and hear. Use binoculars to observe the type of feet the birds have and how they use them. Classify their feet types, using Table 1.

INVESTIGATION   How Does Insulation Affect Body Temperature?

Most mammals maintain a constant body temperature regardless of the temperature of the environment. They produce heat through metabolic processes and maintain that heat in a variety of ways. The rate at which a body loses heat is proportional to the difference between the body's temperature and that of the environment. Therefore, reducing heat loss is essential for mammals that live in environments colder than their bodies. Some mammals control heat loss by limiting the blood supply to the surface of the body. Other mammals produce insulating tissues like fur or blubber that reduce heat loss. For example, the arctic fox has a wooly under-fur and long guard hairs that enable it to exist comfortably at temperatures as low as –40°C.

OBJECTIVES

- Make hypotheses to predict how insulated and uninsulated objects will cool under different temperature conditions.
- Determine the effect of insulation on the cooling of a warm object at room temperature.
- Determine the effect of insulation on cooling of a warm object under cold conditions.
- Compare cooling rates of warm objects at room temperature to the cooling rates of warm objects at low temperature.

MATERIALS

5-L pail
rubber bands (8)
colored pencils
   (4 colors)
hot tap water
   (45–55°C)
masking tape
100-mL graduated
   cylinders (4)
food-storage-size
   plastic bags (4)

ice cubes
   (24 standard cubes)
newspaper (20 pages)
thermometer, 30-cm long
graph paper
stirring rod
clock or stopwatch
laboratory apron
goggles

PROCEDURE

1. Make a **hypothesis** that predicts the effect of insulation on the cooling rate of a warm object. Write your hypothesis in the space provided.

2. Make a second **hypothesis** that describes the effect of surrounding temperature on the cooling rate of a warm object. Write your hypothesis in the space provided.

3. Measure the air temperature of the room and record it in Table 1.

4. Fold 10 pages of newspaper in half crosswise as shown in Figure 1. Fold another 10 pages of newspaper in the same manner.

5. Place 24 ice cubes in the pail and add cold tap water. The water level should be about half the height of the graduated cylinders. Stir gently with a stirring rod until the temperature throughout the pail is the same. Measure the temperature of the ice water in the pail and record it in Table 1 under "Low temperature."

Figure 1

6. Wrap two of the 100-mL graduated cylinders tightly with the folded newspapers as shown in Figure 2. Keep the fold at the base of the cylinder as you wrap. Fasten each wrapped cylinder with two rubber bands. These will be the two insulated cylinders.

How Does Insulation Affect Body Temperature?

PROCEDURE continued

7. Place all four cylinders into separate plastic bags. Fasten each bag around the top of the cylinder with a rubber band as in Figure 3. The cylinders should remain open at the top to allow for insertion of a stirring rod and a thermometer. Label the four bags with your name.

8. Add 100 mL of hot tap water (45–55°C) to each of the four graduated cylinders.

9. Measure the temperature of the water in each of the four graduated cylinders.

10. Place one wrapped and one unwrapped graduated cylinder into the pail of ice water as shown in Figure 4. Leave the other two cylinders at room temperature.

You now have four graduated cylinders containing hot water. These cylinders represent warm-blooded animals. Two animals (cylinders) are placed in cold surroundings (ice water pail), but one animal has body insulation (paper wrap) and one does not. Two animals are placed in room-temperature surroundings, but one animal has body insulation (paper wrap) and one does not.

11. After 5 minutes, measure the water temperature in each of the graduated cylinders and record the data in Table 1. Stir the water in each cylinder with the stirring rod for 15 seconds before inserting the thermometer and reading the temperature.

12. Repeat step 11 every 5 minutes until 30 minutes have elapsed. Record your temperatures each time in Table 1.

13. Make a graph of the data in Table 1 by plotting temperature against time. Time should be plotted on the horizontal axis and temperature on the vertical axis. Plot the data for each cylinder using a different colored pencil.

14. Complete the additional calculations requested in Data and Observations.

HYPOTHESIS 1

_____

_____

_____

_____

Figure 2

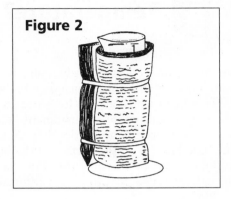

Figure 3

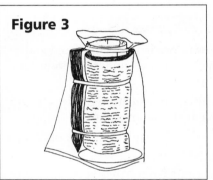

Figure 4

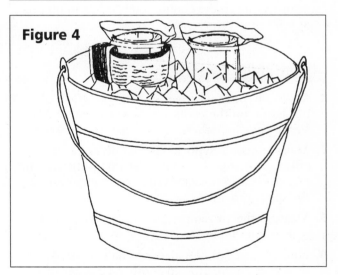

HYPOTHESIS 2

_____

_____

_____

_____

How Does Insulation Affect Body Temperature?

DATA AND OBSERVATIONS

Table 1

Temperature (°C) of Water in Graduated Cylinders				
	Room temperature _____ °C		Low temperature _____ °C	
Time	Uninsulated	Insulated	Uninsulated	Insulated
0 minutes (start)				
5 minutes				
10 minutes				
15 minutes				
20 minutes				
25 minutes				
30 minutes				

1. Calculate the difference between room temperature and the temperature in the uninsulated graduated cylinder at the beginning of the experiment. _____

2. Calculate the difference between the temperature of the ice water and the temperature in the uninsulated graduated cylinder at the beginning of the experiment. _____

Make the following calculations and write your answers in Table 2.

For each cylinder, calculate the temperature difference between starting and ending temperatures.

For each cylinder, calculate the percentage decrease in temperature.

$$\% \text{ decrease in temperature} = \frac{\text{temperature difference}}{\text{temperature at start}}$$

Table 2

Graduated Cylinder	Temperature Difference	% Decrease in Temperature
Uninsulated at room temp.		
Uninsulated at low temp.		
Insulated at room temp.		
Insulated at low temp.		

Lab 32-1

How Does Insulation Affect Body Temperature?

ANALYSIS

1. Examine your graph.

a. Which graduated cylinder lost heat most rapidly? _____

b. Which graduated cylinder lost heat least rapidly? _____

c. What effect did insulation have on the graduated cylinder at room temperature?

_____

d. What effect did insulation have on the graduated cylinder at low temperature?

_____

2. For which of the graduated cylinders was the temperature difference between the contents of the cylinder and the environment greatest at the beginning of the experiment?

_____

3. What effect does environmental temperature appear to have on a body's rate of cooling?

_____

4. How do the percentages of temperature decrease compare between insulated and uninsulated cylinders

a. at room temperature? _____

b. at low temperature? _____

5. Hypothesize how insulation works to maintain heat in a warm body.

_____

6. What other adaptations, in addition to insulation, do mammals use to maintain their body heat?

_____

CHECKING YOUR HYPOTHESIS

Were your **hypotheses** supported by your data? Why or why not?

_____

_____

_____

_____

FURTHER INVESTIGATIONS

1. Conduct this same experiment to compare newspaper with other insulating materials, such as cotton or wool socks.

2. Conduct the same experiment with the same amount of water in cylinders or beakers that have less surface area than the graduated cylinders used in this Investigation. Compare your results.

Mammal Teeth

A beaver with sharp chisel-like teeth gnaws the bark of young aspen trees. A cow collects grasses by holding the plants between its lower teeth and a pad in its upper jaw and tearing off the blades of grass with a jerk of its head. A rat is captured by a weasel with a piercing bite to the back of the rat's neck. A whale strains plankton from sea water with fibrous plates that contain frayed hairlike strands. Mammals have teeth adapted for obtaining specific types of food. By examining a mammal's teeth, you can learn a great deal about its feeding habits and lifestyle.

OBJECTIVES

- Observe adaptations of mammals' teeth for eating particular types of food.
- Identify different types of mammals by the shapes and sizes of their teeth.
- Predict the types of teeth of a mammal based on its feeding behaviors.

PROCEDURE

Part A. Mammal Teeth

1. Examine the following figures of mammal teeth. Compare the descriptions with their matching drawings.

   Figure 1. Evolutionary changes of the cusps, which are the grinding surfaces of teeth, reflect changes in the way teeth are used.

   A. Mammal-like reptile—Fossil evidence shows that the reptilian ancestors of mammals had teeth that probably interlocked when the jaw closed. The teeth may have been adapted for piercing the hard coverings of insects.

   B. Early mammal—Some early mammals may have had triangular teeth that could rub against one another, the beginning of a grinding ability.

   C, D. Later mammals—The change from piercing types of teeth to teeth with larger areas of touching surfaces enabled mammals to cut and grind their food.

   Figure 2. Mammal tooth types—Most mammals have a variety of types of teeth. Incisors gnaw or cut, canines stab and tear, premolars crush and shear, and molars grind.

Figure 1

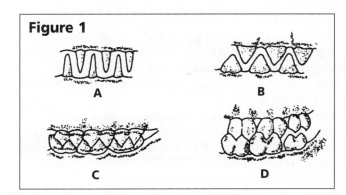

Figure 2

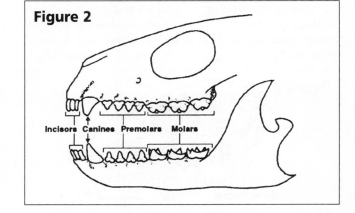

Mammal Teeth

PROCEDURE continued

Figure 3. Insectivores—In insectivores, such as shrews and moles, the incisors are simple and peg-like. The canines are usually the same shape as the incisors. The molars and premolars have sharp points suited for shearing and cutting through hard insect coverings.

Figure 4. Bats are varied in their food preferences.

A. Fruit-eating bats—Teeth are specialized for crushing small, hard fruits.

B. Nectar-feeding bats—Teeth are very small.

C. Insect-feeding bats—Teeth are all very sharp and can easily capture and pierce insects with hard coverings.

Figure 5. Anteaters—Anteaters have no teeth or only a few tiny teeth. They use their tongues to catch ants and termites. They expose the insects by pulling apart the insect nests with their powerful front arms.

Figure 6. Rodents—Rodents, such as the beaver, are equipped with large incisors adapted to gnawing trees. The teeth continue to grow throughout the rodent's life. The sharp chisel-like edge is maintained by one side wearing away more quickly than the other. The molars grind very tough bark.

Figure 7. Carnivores—Carnivores, such as the coyote, have long canines, shearing premolars, and crushing molars. These teeth enable them to capture and kill their prey.

Figure 8. Omnivores—Omnivores, such as the black bear, often have large canines for tearing and large molars for grinding. Premolars and incisors are not extremely large and sharp. In the case of this type of bear, they are greatly reduced.

Figure 9. Ungulates—Ungulates, such as the pronghorn antelope, are browsers or grazers. Incisors are adapted for nipping off grasses and shrubs. Canines are absent. Premolars and molars are long in young animals and short in older animals that have worn them away by grinding coarse plant material.

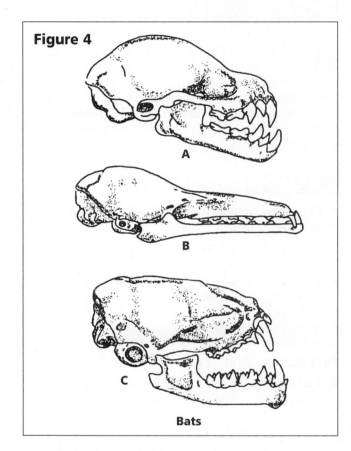

Figure 4

Bats

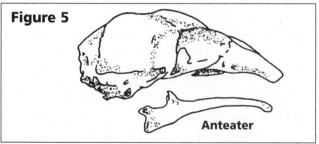

Figure 5

Anteater

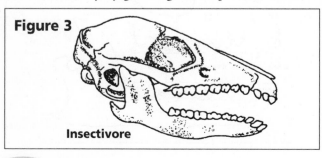

Figure 3

Insectivore

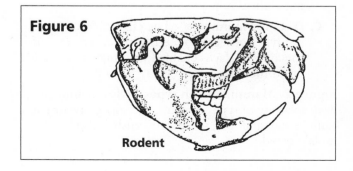

Figure 6

Rodent

Lab

Mammal Teeth

32-2

Figure 7

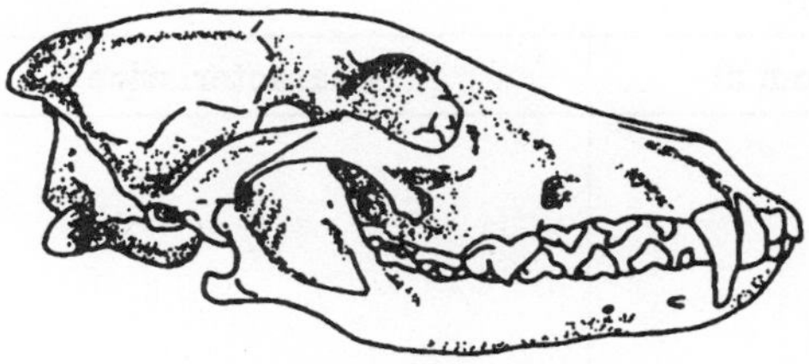

Carnivore

Figure 8

Omnivore

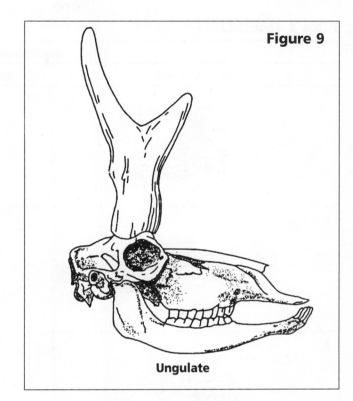

Figure 9

Ungulate

1. Examine the mammal skulls in Table 1 and compare their teeth with the teeth in Part A.

2. Identify the type of mammal, illustrated in figures 3–9, represented by each set of teeth and write its name next to the skull.

3. Write the teeth characteristics you used to identify each mammal type.

DATA AND OBSERVATIONS

Table 1

Skull Identification		
Mammal skull	Type of mammal	Characteristics
1. 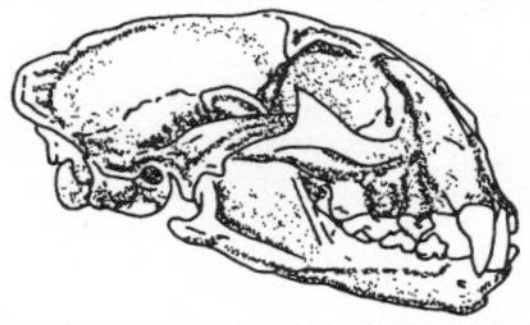		
2. 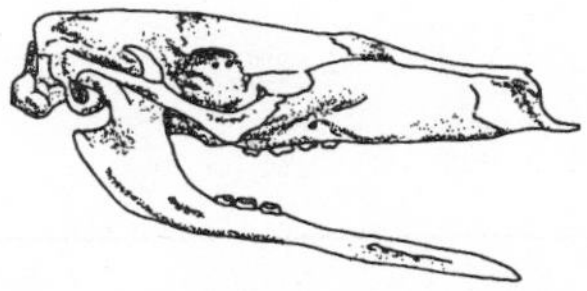		

Mammal Teeth

Table 1 (continued)

Skull Identification		
Mammal skull	**Type of mammal**	**Characteristics**
3.		
4.		
5.		
6.		
7.		

Mammal Teeth

Table 1 (continued)

Skull Identification		
Mammal skull	**Type of mammal**	**Characteristics**
8.		
9.		
10.		

ANALYSIS

1. Based on your observations of the teeth in Table 1, tell whether each of these ten mammals eats plants, insects, meat, or a variety of foods.

1. _____ 4. _____ 7. _____ 10. _____

2. _____ 5. _____ 8. _____

3. _____ 6. _____ 9. _____

2. Based on your answers to question 1, hypothesize whether each mammal moves quickly or slowly.

_____

_____

_____

_____

ANALYSIS continued

3. Compare the differences among the teeth of omnivores, carnivores, and herbivores.

_____

_____

4. Figure 10 is the skull of a vampire bat. This bat makes an incision in the skin of a cow or horse and laps the blood with its tongue. Explain how the teeth of a vampire bat are adapted to this behavior.

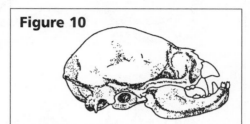

Figure 10

_____

_____

_____

5. Figure 11 is the skull of a walrus. Walruses feed on mollusks that they rake from the sea floor. Explain how the teeth of a walrus are adapted to this lifestyle.

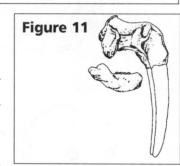

Figure 11

_____

_____

6. Describe the teeth of a mammal that forages on the floor of the forest for worms.

_____

_____

7. If a mammal fed on only mosses and lichens, what would its teeth be like?

_____

_____

8. If a sheep were a predator, how would its teeth be different?

_____

_____

FURTHER EXPLORATIONS

1. Take a field trip to a museum of natural history and examine the teeth of as many mounted mammals as you can. Identify the mammals or herbivores, omnivores, or carnivores based on their teeth. Use library references to verify the eating habits of the mammals.

2. Do research to examine the differences between molars, premolars, canines, and incisors. For each tooth, make a diagram that identifies the parts of the teeth: enamel, dentine, and pulp cavity.

INVESTIGATION

How Is Response Related to Nervous System Complexity?

The movement of an organism toward or away from an environmental stimulus is called a taxis. Such an ability might mean the difference between life and death for an organism. Organisms that respond instinctively to stimuli do not have to learn the correct response because all the behavior necessary for survival is programmed in their genetic material. Each pairing of stimulus and response is an innate behavior that contributes to the survival of the organism. Responses of different invertebrates to the force of gravity (geotaxis), chemicals (chemotaxis), and light (phototaxis) can be compared.

OBJECTIVES

- Hypothesize the effect of nervous system complexity on responses to stimuli.
- Compare the responses of different invertebrates to the same stimuli.

MATERIALS

droppers (2)
flashlight
coarse salt
black construction paper
scissors

plastic thermometer
    tube with caps
living cultures of:
    planarians
    vinegar eels
    (*Turbatrix aceti*)
    Daphnia

filter paper (3 pieces)
duct tape
test-tube rack
wax marking pencil
ruler

stopwatch
hand lens
laboratory apron
goggles

PROCEDURE

The three species of invertebrates in this Investigation have different levels of nervous system development. Planarians have the least developed nervous system. *Daphnia* has the most developed nervous system.

1. Make a **hypothesis** that predicts how a species' response to a stimulus relates to the complexity of its nervous system. Write your hypothesis in the space provided.

2. Place a cap tightly on one end of a thermometer tube. Fill the tube with water to test for leakage. If the cap leaks, tape it securely with a piece of duct tape as shown in Figure 1 so that it is watertight. Pour out the water. Use the ruler and marking pencil to divide the tube into thirds as shown in Figure 1.

3. Using a clean dropper, fill the tube with the planarian culture medium. Be sure the tube contains at least three organisms.

4. Place the second cap on the top of your tube and tape the cap if it leaks. Place the tube in the test-

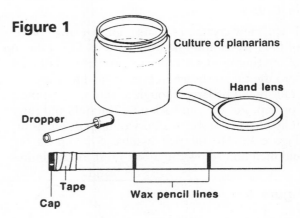

Figure 1

Culture of planarians

Hand lens

Dropper

Tape

Cap

Wax pencil lines

tube rack as shown in Figure 2. Immediately start your stopwatch and measure the amount of time it takes for you to see that the planarians are reacting to the force of gravity. Stop your stopwatch when you detect an overall trend in movement in the tube. Record this time in Table 1 in the "Geotaxis" section.

How Is Response Related to Nervous System Complexity?

PROCEDURE continued

5. After 5 minutes, observe the planarians and look for their general direction of movement. Count the number of planarians in the top third of the tube and the number in the bottom third of the tube. Do not count the planarians in the middle third. Use a hand lens to help you see the planarians. Record these numbers in the "Geotaxis" section in Table 1.

6. Place the tube containing the planarians in a horizontal position. Cover one end with a piece of black construction paper and arrange a flashlight so that it clearly illuminates the other end of the tube as shown in Figure 3. Immediately start your stopwatch and measure the amount of time it takes for the planarians to react to light. Stop your stopwatch when you detect an overall trend in movement in the tube. Record this time in Table 1 in the "Phototaxis"'s section.

7. After 5 minutes, observe the planarians and note the general direction of movement. Remove the construction paper and flashlight. Count the number of planarians in the lighted third of the tube and the number in the dark third. Do not count planarians in the middle third of the tube. Record these numbers in the "Phototaxis" section in Table 1.

8. Place a few crystals of coarse salt in a small piece of filter paper and fold it into a wad. Open the top of the thermometer tube and put the wad of paper into the cap as shown in Figure 4a. The paper wad should be large enough to stay stuck in the top cap. Put the cap back onto the tube and carefully place the tube in a horizontal position as in Figure 4b. The salt will dissolve slowly in the liquid medium. Allow the tube to rest for 2 minutes.

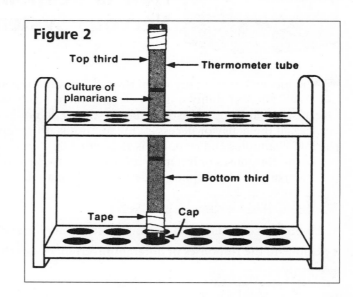

Figure 2

Top third
Thermometer tube
Culture of planarians
Bottom third
Tape
Cap

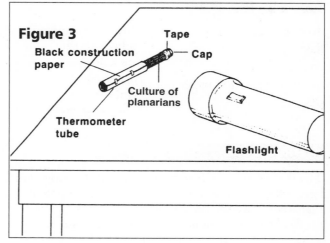

Figure 3

Tape
Black construction paper
Cap
Culture of planarians
Thermometer tube
Flashlight

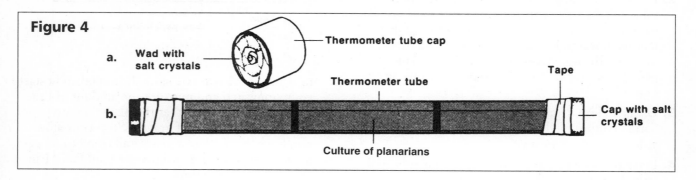

Figure 4

a.
Wad with salt crystals
Thermometer tube cap

b.
Thermometer tube
Tape
Cap with salt crystals
Culture of planarians

How Is Response Related to Nervous System Complexity?

9. Start your stopwatch and measure the reaction time as described in steps 4 and 6. Record the time in the "Chemotaxis" section in Table 1.

10. After 5 minutes, observe the planarians and note their general direction of movement. Count the number of planarians in the third of the tube nearest the salt and the number in the third farthest from the salt. Record these numbers in the "Chemotaxis" section in Table 1.

11. Complete Table 1 for planarians by entering their responses to the stimuli. Follow the directions below Table 1.

12. Dispose of the planarians in your tube according to your teacher's directions. Clean and rinse the tube.

13. Repeat steps 3 through 12 for the other two cultures of organisms.

HYPOTHESIS

_____

_____

_____

DATA AND OBSERVATIONS

Table 1

Species' Responses to Stimuli				
Response		**Planarians**	**Vinegar eel**	*Daphnia*
Geotaxis	Reaction time (min)			
	Number at top			
	Number at bottom			
	Response (+, −, 0)			
Phototaxis	Reaction time (min)			
	Number at light end			
	Number at dark end			
	Response (+, −, 0)			
Chemotaxis	Reaction time (min)			
	Number near salt			
	Number away from salt			
	Response (+, −, 0)			

If number of organisms close to the stimulus is much greater than number away from the stimulus, the response is strongly positive (+).

If number of organisms close to the stimulus is much less than number away from the stimulus, the response is strongly negative (−).

If number of organisms close to the stimulus is about equal to number away from the stimulus, the response is weak or absent (0).

How Is Response Related to Nervous System Complexity?

ANALYSIS

1. Which species exhibited:

positive geotaxis? _____ negative geotaxis? _____

positive phototaxis? _____ negative phototaxis? _____

positive chemotaxis? _____ negative chemotaxis? _____

2. In which species did the individual organisms exhibit the most diverse response to:

the force of gravity? _____

light? _____

salt? _____

3. Is diversity of response among the individuals of a species a sign of a greater or less complex nervous system?

_____

Explain. _____

_____

4. a. Which species had the fastest reaction time? _____

b. Why would this automatic, rapid response to a stimulus be an adaptive advantage in a simple organism?

_____

5. What might be some reasons for a species not demonstrating a strongly positive or negative response?

_____

_____

CHECKING YOUR HYPOTHESIS

Was your **hypothesis** supported by your data? Why or why not?

_____

_____

FURTHER INVESTIGATIONS

1. Design an experiment to show whether or not invertebrates can learn. Design a simple maze and use rewards, such as food, to test an invertebrate's ability to learn.

2. More complex organisms can experience innate responses similar to taxes, but often overcome them. For example, your natural response to a pin prick would be to move away, but many people give themselves injections for medical purposes. Investigate the parts of the mammalian brain, including those responsible for simple instinctive responses and those that allow higher learning.

Conditioning in Guinea Pigs

Animals respond to stimuli in their surroundings. Usually each stimulus causes a specific response. An animal can be trained to give an established response to a new stimulus when the new stimulus is given at the same time as the original stimulus. This procedure is repeated until the animal learns to respond to the new stimulus when it is given alone. This kind of training is called conditioning because the animal learns to respond to a new set of conditions. The learned response to a new stimulus is called a conditioned response.

OBJECTIVES

- Observe feeding behavior in a guinea pig.
- Carry out an experiment to produce a conditioned response in a guinea pig.
- Determine the amount of time needed to produce a conditioned response in a guinea pig.

MATERIALS

guinea pig with cage
  and bedding
pellet-type food
water
lettuce, carrots, apples,
  or other fresh food

bell, whistle, or other
  source of sound
laboratory apron
goggles

PROCEDURE

1. Obtain a guinea pig; be sure that it has an adequate cage and appropriate bedding materials as shown in Figure 1.

2. Become familiar with the nutritional needs of the guinea pig. A guinea pig needs fresh water and between 40g and 70g of food daily. Table 1 lists some foods that are appropriate for a guinea pig. If the guinea pig receives a high-protein, high

carbohydrate food, less food is needed than if the diet is lower in proteins and carbohydrates. Although many of the dry foods contain less protein and carbohydrate, they are important in the animal's diet because they contain a variety of essential vitamins.

3. Vary the guinea pig's diet until you determine which combination of foods it prefers.

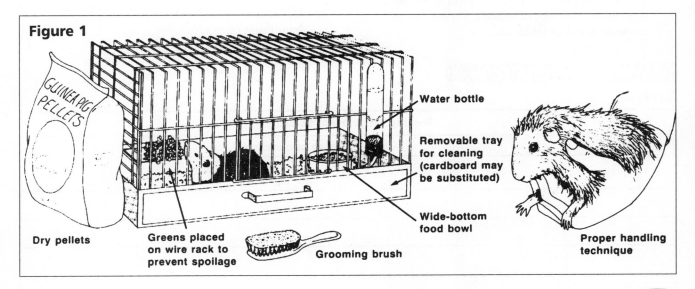

Figure 1

Water bottle

Removable tray
for cleaning
(cardboard may
be substituted)

Wide-bottom
food bowl

Dry pellets

Greens placed
on wire rack to
prevent spoilage

Grooming brush

Proper handling
technique

Conditioning in Guinea Pigs

PROCEDURE continued

Table 1

Food	Percent protein	Percent carbohydrate
Corn	9	65
Peanuts	20	15
Dry whole-wheat bread	8	55
Guinea pig pellets	20	48
Alfalfa hay	4	10
Lettuce	2	8
Carrots	1	8
Turnips	0.7	9
Apples	0.2	12
Pears	0.5	14

4. As you prepare the food, have a classmate observe the animal's behavior. Guinea pigs have very good hearing, and they may associate the sound of food preparation with being fed.

5. When you feed the guinea pig, watch its behavior closely. See if you can determine the point at which it recognizes that it is about to be fed. Record your observations in Table 2. Guinea pigs make a variety of sounds. Be sure to include these in the data table, too.

6. Observe the guinea pig when you approach the cage but do not feed it. Record this non-feeding behavior in Table 3.

7. After five days of observations, study tables 2 and 3. Look for patterns of behavior that are associated only with feeding. These will be the behaviors that you will look for to determine when the guinea pig has been conditioned.

8. Just before you feed the guinea pig, provide a new stimulus. Make a loud hand clap or another sound. You may wish to try a visual stimulus of some kind. Make sure that the stimulus is one that does not occur at any other time.

9. Continue to condition the guinea pig to respond to the new stimulus for one week. In Table 4, record the animal's behavior. As you do this, be aware of any other stimuli that might be associated with feeding. You will need to remove them before you can test the effects of conditioning.

10. After one week, give the new stimulus but do not feed the guinea pig immediately. Look for behavior related to feeding. Then feed the animal. Record your observations in Table 4 next to "Testing."

11. If the guinea pig did not show feeding behavior in response to the new stimulus, repeat the conditioning process for another week.

12. After a week, test the conditioning again.

13. Repeat steps 11 and 12 until the guinea pig responds to the new stimulus with behavior that is related to feeding.

14. Wash your hands with soap and water after handling your guinea pig.

DATA AND OBSERVATIONS

Table 2

Feeding Behaviors of Guinea Pig	
Date	Behavior observed

Conditioning in Guinea Pigs

Table 3

Non-Feeding Behaviors of Guinea Pig	
Date	Behavior observed

Table 4

Conditioning of Guinea Pig		
Date	Procedure	Behavior observed
	Conditioning	
	Conditioning	
	Conditioning	
	Conditioning	
	Conditioning	
	Conditioning	
	Conditioning	
	Testing	
	Conditioning	
	Conditioning	
	Conditioning	
	Conditioning	
	Conditioning	
	Conditioning	
	Conditioning	
	Testing	

Conditioning in Guinea Pigs

ANALYSIS

1. What behavior was seen only when the guinea pig expected to be fed?

_____

_____

2. What did you choose as the new stimulus?

_____

_____

3. Why did the new stimulus have to be one that would not occur at any other time?

_____

_____

4. Why did other stimuli related to feeding have to be removed before the conditioning could be tested?

_____

_____

5. How long did it take to condition the guinea pig to the new stimulus?

_____

_____

6. How did the guinea pig respond to the new stimulus?

_____

_____

FURTHER EXPLORATIONS

1. Stop using the new stimulus for a week. Then test it again. Determine whether the conditioned response appears. Explain your observation.

2. Research Pavlov's experiments in conditioning. Make particular note of the way he controlled extra stimuli in his subjects' environments. Write a report on your research.

EXPLORATION   The Skeletal System

The skeletal system of most vertebrates is composed of bones like those that make up your own skeleton. Bones form as minerals are deposited around special bone-forming cells found in cartilage. As the bone forms, pathways for blood vessels and nerves are established within the mineral network.

OBJECTIVES

- Observe a microscope slide of compact bone and identify the different structures.
- Identify bones by their shapes.
- Observe the features of joints.

MATERIALS

prepared slide of compact bone

compound light microscope

model of human skeleton or picture of human skeleton

PROCEDURE

Part A. Bone Cells

1. Obtain a slide of compact bone and use the 10× objective to focus on the tissue.

2. Compact bone is made up of many osteon systems, elongated structures that surround canals. In cross section, an osteon system looks like a series of rings, each surrounding a canal and containing osteocytes, or bone cells. Look at the osteon systems shown in Figure 1 and locate similar structures on your slide.

3. An osteon canal is located in the center of each system. Locate an osteon canal on your slide. Blood vessels and nerves are found in each canal.

4. Osteocytes are embedded in mineral salts in layers around osteon canals. Find these layers on your slide. Within each layer are small canals called canaliculi that carry fluids between the blood vessels and osteocytes.

5. In Data and Observations, draw an osteon system as observed on your slide. Label an osteon canal, osteocytes, and canaliculi.

Figure 1

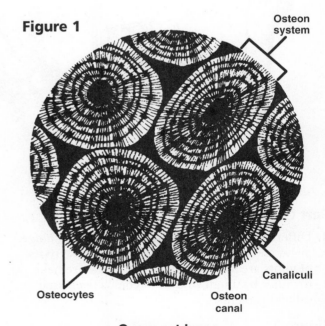

Compact bone

Lab 34-1

The Skeletal System

PROCEDURE continued

Part B. Recognizing Bones

1. Each bone has its own shape and function. Compare the human bones shown in Figure 2 with a model of a human skeleton or a picture of a human skeleton.

2. For each bone, determine where it is found in the human body. In Table 1, record the location, scientific name, and common name of each bone. Use library references if necessary. **NOTE:** *Observe the ends of the bones to help you distinguish them.*

Figure 2

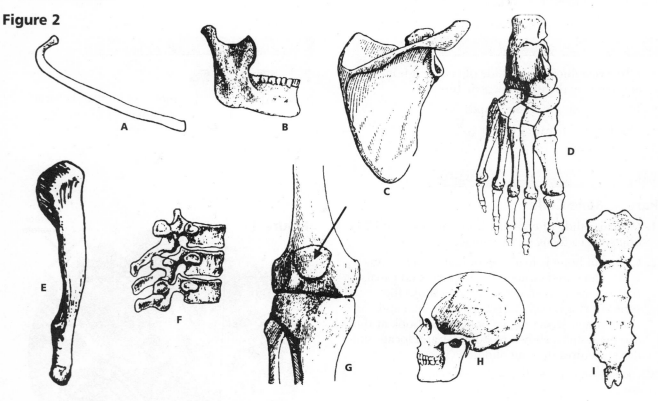

Part C. Joints

1. Different joint types include fixed, gliding, hinge, pivot, and ball-and-socket. Three of these types of joints are shown in Figure 3. Joints are held together by tough strands of connective tissue called ligaments.

2. Study the examples of joints in Figure 3. Identify the type of joint shown in each drawing. In Table 2, record the type of joint shown and its location in the body.

Figure 3

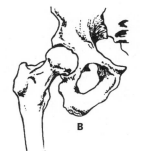
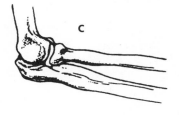

The Skeletal System

DATA AND OBSERVATIONS

Table 1

Bone	Common Name of Bone(s)	Scientific Name of Bone(s)	Location in Body
A			
B			
C			
D			
E			
F			
G			
H			
I			

Table 2

Joint	Type	Location in Body
A		
B		
C		

The Skeletal System

ANALYSIS

1. In Part B, you examined various bones. What features did you use to distinguish the various bones from one another?

_____

_____

2. Which type of joint allows only for circular movement? Give an example.

_____

3. Which type of joint allows the bones to move back and forth? Give an example.

_____

4. Which type of joint allows bones to twist around each other?

_____

5. What is the function of the:

a. osteon canal? _____

b. canaliculi? _____

FURTHER EXPLORATIONS

1. If individual human bones or models of human bones are available, identify each bone by common name and by scientific name. Describe what features you used to identify the bones.

2. If a skeleton of another vertebrate, such as a cat, is available, identify the major bones of the skeleton. If such a skeleton is not available, do library research to find a drawing of a vertebrate skeleton. Make a table to list the names of bones found both in this skeleton and in the human skeleton. In your table, list ways these bones are similar and ways they are different.

How Much Vitamin C Are You Getting?

Vitamin C is an essential vitamin for human health. Since the body cannot manufacture or store this chemical, vitamin C must be part of one's daily diet. It is found in potatoes as well as in a variety of citrus fruits and can also be obtained from a vitamin supplement. A daily intake of 60 milligrams, or 0.060 grams, is considered the minimum daily requirement for adults, as determined by the United States Food and Drug Administration.

The vitamin C concentrations of various citrus juices can be determined by comparing them to a standard vitamin C solution of known concentration. The microchemistry technique used to make the comparison is called a titration. In this experiment, titration is first used to measure the volume of iodine solution needed to react completely with a known volume of the standard vitamin C solution. Once that volume of iodine solution is determined, it is then possible to titrate and analyze unknown concentrations of vitamin C. If, for example, it requires 10 drops of iodine solution to completely react with 5 milligrams of a standard vitamin C solution, then an unknown solution requiring 20 drops of iodine solution must contain twice as much vitamin C as the standard solution.

The iodine solution is added to the standard vitamin C solution one drop at a time. A chemical reaction occurs, during which the brown-red color of the iodine disappears. The end point of the reaction is reached when sufficient iodine is added to react with all the vitamin C. The end point can be detected by a starch indicator, which is added to the standard solution at the beginning of the experiment. The first drop of excess iodine causes the starch solution to turn blue-black, indicating that all of the vitamin C has been used up in the reaction.

OBJECTIVES

- Measure the volume of iodine solution needed to react with a standard concentration of vitamin C.
- Hypothesize which citrus-juice sample contains the most vitamin C per serving.

- Measure the volume of iodine solution needed to react with citrus-juice samples containing unknown concentrations of vitamin C.
- Calculate the concentration of vitamin C in the citrus-juice samples.

MATERIALS

standard vitamin C
    solution (1 mg/mL)
1% starch solution
distilled water
iodine solution

citrus-juice samples
    (4 kinds, including
    fresh orange juice)
paper towels
plastic pipettes or
    microtip pipettes
    (4 each)

toothpicks (5)
microplate, 96-well
microplate, 24-well
scissors
hole punch
plastic straw

wax marking pencil
laboratory apron
goggles

How Much Vitamin C Are You Getting?

PROCEDURE

Part A. The Standard Solution

1. Construct a ring stand and pipette holder using a plastic drinking straw. With scissors, cut the straw into two pieces on a slant, as shown in Figure 1. Insert the uncut end of one piece into a well on the edge of a 96-well microplate. This piece is a ring stand.

2. Cut off the slanted end of the other piece of straw, as shown in Figure 1. Insert one blade of the scissors into the straw and make a length-wise cut to about the halfway point of the straw. Make a similar cut opposite and parallel to the first cut. Join the cuts by making a crosswise cut. Discard the piece of straw that results.

3. Cut the end of the flap of straw created in step 2 to make a point. Bend the flap and insert its point into the opening of the uncut portion of the straw, as shown in steps 3 and 4. The loop you form in this way will serve as a pipette holder.

4. Hold the pipette holder with the loop horizontal and use a hole punch to make a hole at the end of the straw opposite the loop. Slip the pipette holder onto the ring-stand straw in the microplate.

5. Fill one of the pipettes with iodine solution. Insert the pipette into the ring stand and position the pipette directly over a well in the 24-well microplate. (See Figure 1.) This pipette will be used to titrate the standard solution and then (in Part B) the solutions of unknown vitamin C concentration. **CAUTION:** *Iodine is a poison. If spillage occurs, rinse with water and call the teacher immediately.*

6. Use a second pipette to place 10 drops of the standard vitamin C solution in the well of the 24-well microplate under the pipette.

Figure 1

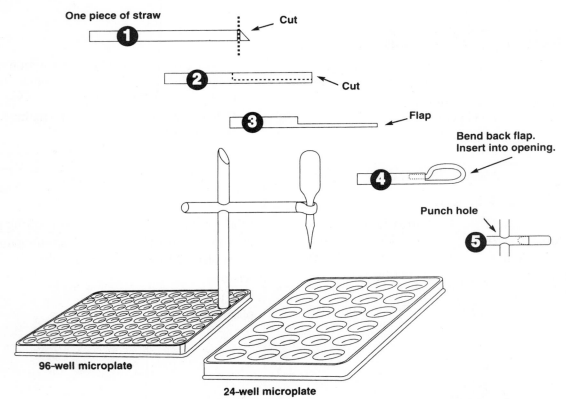

How Much Vitamin C Are You Getting?

7. Use a separate pipette to place 10 drops of water and 5 drops of starch solution in the same well. Mix the contents of the well with the toothpick.

8. Now you are ready to begin the titration. Add the iodine solution to the well one drop at a time, stirring gently with the toothpick and counting each of the drops. When the brown-red color of the iodine begins to disappear, add the iodine solution more slowly. When the color has completely disappeared, the next drop or two may signal the end point of the chemical reaction. As soon as you see a blue-black tint, stop the titration and record in Table 1 the total number of drops of iodine you used.

9. Discard the contents of the well according to the teacher's directions. Rinse the well with distilled water and dry it with a paper towel.

10. Repeat steps 5–9 two more times. Record each trial in Table 1 and average your results.

11. Rinse the stem and bulb of the pipette from the standard vitamin C in preparation for its use in Part B.

Part B. Unknown Solutions

1. Label the citrus-juice samples A–D.

2. Make a **hypothesis** to predict which juice sample has the highest vitamin C content. Write your hypothesis in the space provided.

3. If necessary, refill the pipette in the ring stand with iodine solution.

4. Add 10 drops of juice A to a well in the 24-well microplate.

5. Add 10 drops of water and 5 drops of starch solution to the same well. Mix the contents of the well with a toothpick.

6. Add the iodine solution to the well drop by drop, stirring steadily and counting each of the drops.

7. Record in Table 1 the number of drops required to turn the solution blue-black.

8. Discard the contents of the well according to the teacher's directions. Rinse the well and dry it with a paper towel.

9. Repeat steps 3–8 two more times and average your results.

10. Rinse the stem and bulb of the pipette from the juice sample.

11. Repeat steps 3–10 for samples B–D.

12. Dispose of all materials and clean all glassware as directed by your teacher. Wash your hands thoroughly.

HYPOTHESIS

_____

_____

DATA AND OBSERVATIONS

Table 1

	Trial 1	Trial 2	Trial 3	Average
Drops of Iodine Solution (Std. Vit. C)				
Drops of Iodine Solution (Juice A)				
Drops of Iodine Solution (Juice B)				
Drops of Iodine Solution (Juice C)				
Drops of Iodine Solution (Juice D)				

How Much Vitamin C Are You Getting?

ANALYSIS

1. If your class has worked in teams, collect and share the data. For each kind of citrus juice, calculate the ratio of drops of iodine added to the juice to the drops of iodine added to the standard solution as shown below. Record the ratios in Table 2.

$$\text{Ratio for juice} = \frac{\text{Average \# of drops added to juice}}{\text{Average \# of drops added to standard solution}}$$

2. For each juice, calculate the concentration of vitamin C in mg per mL by multiplying the ratio of iodine drops for that juice by the concentration of vitamin C in the standard solution (1 mg/mL), as shown below. Record the concentrations in Table 2.

Concentration of vitamin C in juice = Ratio for juice × 1 mg/mL

3. As shown below, calculate the number of milligrams of vitamin C you would get by drinking an average serving of each juice. (The average serving size of citrus juice is 150 mL.) Record the number of milligrams in Table 2.

\# of mg of vitamin C in 1 serving of juice = Concentration of vitamin C in juice × 150 mL

Table 2

Juice Sample	Ratio	Concentration of Vitamin C (mg/mL)	Amount of Vitamin C in 1 serving (mg)
A			
B			
C			
D			

4. What would be another way of calculating the amount of iodine solution used in each titration, besides counting drops?

_____

_____

CHECKING YOUR HYPOTHESIS

Was your **hypothesis** supported by the class data? Why or why not?

_____

_____

FURTHER INVESTIGATIONS

1. Design an experiment to assay the vitamin C content of potatoes.
2. Design an experiment to assay a multivitamin tablet for vitamin C content.

Caloric Content of a Meal

Food is the source of energy and building materials for the human body. The energy is used by cells to carry out respiration, protein synthesis, active transport, and other metabolic processes as a person goes about his or her daily activities. When the foods a person eats provide more energy than is needed, the excess is converted to fat and stored in the body. Conversely, when the body receives less energy from food than it needs, the energy in stored fat is released.

OBJECTIVES

- Calculate the number of Calories and grams of carbohydrate, fat, and protein in two meals.
- Appraise the nutritional value of each meal.
- Plan a balanced, nutritional meal.

MATERIALS

food table with Calories and grams of carbohydrate, fat, and protein in each food listed

PROCEDURE

1. Study Table 1. Assume a person eats three meals per day. Calculate the recommended intake *per meal* for all of the nutrients and Calories in Table 1. Record these in Table 1.

2. Examine the list of foods in the first column of Table 2. Two separate lunch plans are presented. Provide the information for the remaining columns by locating the proper food in the food table. If the food table does not present information for the same serving size, you will have to calculate the correct values. First, calculate the

Calories and grams of nutrients per unit from the food table, and then multiply that number by the correct serving size.

3. Record your information, including the totals for each meal, in Table 2.

4. Compare the information for meals 1 and 2 in Table 2 with the recommended intake per meal in Table 1.

5. Plan a meal that comes close to the intake recommendations of Table 1. Use the food table as a guide.

DATA AND OBSERVATIONS

Table 1

Recommended Food Intake		
Nutrients	**Amount per day**	**Amount per meal**
Calories	2000–2500	
Carbohydrate (grams)	120	
Fat (grams)	60	
Protein (grams)	60	

Caloric Content of a Meal

Table 2

Food	Serving size	Calories	Carbohydrate (grams)	Fat (grams)	Protein (grams)
Calories and Nutrients of Two Sample Meals					
Meal 1					
Spaghetti with meat sauce	6 oz				
Green beans	4 oz				
Italian bread	2 slices				
Butter	1 Tbsp				
Gelatin dessert	4 oz				
Total					
Meal 2					
Hamburger bun	1				
Ground beef	4 oz				
Cheese (American)	1 oz				
Ketchup	2 Tbsp				
French fries	24				
Cola-type beverage	12 oz				
Total					

ANALYSIS

1. Which of the two sample meals in Table 2 is higher in: Calories? _____ fat? _____ carbohydrate? _____ protein? _____

2. Which of the two sample meals is more nutritious? Why?_____

3. Sample meal 1 is low in which important nutrient? _____ What kinds of foods could the other two meals of the day include to make up for this? _____

4. If you eat more Calories than your daily activities use, what will happen to your weight?

_____

5. If you eat fewer Calories than you need, what will happen to your weight? _____

FURTHER EXPLORATIONS

1. Prepare a weekly meal plan that meets the recommended daily intake values of Calories, carbohydrates, fat, and protein shown in Table 1.

2. Record the number of Calories and amounts of nutrients you had for breakfast and lunch. Determine how many Calories and how much of each nutrient you need for the evening meal. Plan this meal so that you will have received 100% of the recommended daily food intake values.

What is the Effect of Smell on Taste?

Chemoreceptors are receptors that respond to chemical stimuli. Chemoreceptors for smell respond to air-borne chemicals; chemoreceptors for taste respond to chemicals in your mouth. Four different types of chemoreceptors in the mouth respond to four different kinds of taste: sweet, sour, bitter, and salty.

OBJECTIVES

- Hypothesize the effect of smell on the ability to identify tastes.
- Determine the effect of smell on the ability to identify tastes.
- Use correct lab techniques to dilute solutions.

MATERIALS

paper cups (10)
permanent marking pen
sweet solution
cotton swabs (10)
water
paper towel
paper bag

25-mL graduated
    cylinder
stirring rod
toothpicks (18)
bread, apple, and orange
    (6 pieces of each)
plastic wrap

PROCEDURE

Part A. The Effect of Smell on Taste

1. Label three paper cups "bread," "apple," or "orange." Place six pieces of each food into the appropriate cup, and cover each cup with plastic wrap. Keep the foods covered during the activity to contain their odors.

2. Form a hypothesis about how the sense of smell affects the sense of taste. Write your hypothesis in the space provided on the next page.

3. Have your partner sit with eyes closed, holding his or her nose. Use a toothpick to spear a piece of one of the foods and place it in your partner's mouth. Your partner should slowly chew and then swallow the food, and then take a sip of water to clear the tongue. Discard the toothpick immediately into the paper bag.

4. Ask your partner to identify the food. If the identification was correct, mark "+" in the "Taste only" column in Table 1. If the identification was incorrect, mark "−" in the "Taste only" column.

5. Repeat steps 3 and 4 with the other two foods. Record all data in Table 1.

6. Have your partner sit with eyes closed, but not holding his or her nose. Use a toothpick to hold a piece of one of the foods beneath your partner's nose. Discard the toothpick and the food piece.

7. Ask your partner to identify each food. Record the responses under the column labeled "Smell only" in Table 1.

8. Have your partner sit with eyes closed, but not holding his or her nose. Repeat steps 3–5. Record the responses under the column labeled "Taste and smell" in Table 1.

9. Reverse roles, and repeat steps 3–8.

10. Discard the food cups in the bag.

Part B. The Threshold of Taste

1. Use a marking pen to label 5 paper cups "full-strength," "1/2-strength," "1/4-strength," "1/8-strength," and "1/16-strength."

2. Measure 10 mL of sweet solution into the cup marked "full-strength."

What is the Effect of Smell on Taste?

PROCEDURE continued

3. Measure 10 mL of sweet solution into the cup marked "1/2-strength." Add 10 mL of water, and stir the mixture. Rinse the graduated cylinder with water. You will dilute this mixture several times. Refer to Figure 1 as you follow the directions for diluting the solution.

4. Measure 10 mL of the 1/2-strength solution into the cup marked "1/4-strength." Add 10 mL of water, and stir the mixture. Rinse the cylinder with water.

5. Measure 10 mL of the 1/4-strength solution into the cup marked "1/8-strength." Add 10 mL of water, and stir the mixture. Rinse the cylinder with water.

6. Measure 10 mL of the 1/8-strength solution into the cup marked "1/16-strength." Add 10 mL of water, and stir the mixture.

7. Obtain 5 cotton swabs. Use them to test your partner's ability to taste the solutions of different strengths. Test the front of the tongue only. Start with the full-strength solution and then test progressively weaker solutions. Immediately discard each swab.

8. Record your partner's ability to taste each solution. If the taste was recognized, mark "+" in the appropriate space in Table 2. If the taste was not recognized, mark "–" in the appropriate space in Table 2. Allow your partner to take a sip of water now and then.

The threshold of taste is the minimum concentration of solution that can be tasted. Your partner's exact threshold is somewhere in between the strength of the weakest solution that could be tasted and the strongest solution that could not be tasted.

9. Reverse roles, and repeat steps 7 and 8. You may reuse the test solutions.

Figure 1

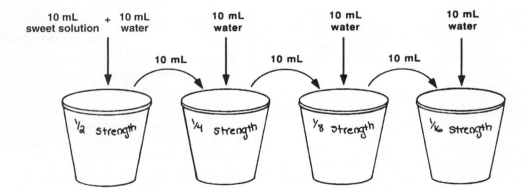

HYPOTHESIS

_____

_____

What is the Effect of Smell on Taste?

Lab 36-1

DATA AND OBSERVATIONS

Table 1

Combining Smell and Taste			
Food	Taste only	Smell only	Taste and smell
Potato			
Apple			
Onion			

Table 2

Threshold of Taste	
Strength of solution	Taste
Full	
1/2	
1/4	
1/8	
1/16	

ANALYSIS

1. Under which conditions did you and your partner make the most accurate identifications of foods?

_____

2. Why do foods seem to have less taste when you have a cold? _____

_____

3. Did you and your partner have the same threshold of taste? _____

4. How could you measure more exactly someone's threshold of taste? _____

_____

5. Explain how damage to the areas of the brain that receive sensory input from olfactory neurons and taste buds might affect a person's sense of smell or taste. _____

_____

_____

What is the Effect of Smell on Taste?

CHECKING YOUR HYPOTHESIS

Was your **hypothesis** supported by your data? Why or why not?

_____

_____

FURTHER INVESTIGATIONS

1. Have students define the terms ageusia, anosmia, anosphrasia, agnosia, and synesthesia.

2. Repeat the threshold of taste test using one of the other solutions. Test several areas of the tongue to find out if the threshold for a given taste is the same for all areas of the tongue.

Design Your Own

INVESTIGATION

What Is the Effect of Exercise on Heart Rate?

Lab

37-1

The heart pumps blood to all the cells of the body. As a person exercises, the muscle cells use increased amounts of oxygen and nutrients, which must be replaced by the blood. The muscle cells also produce more wastes, which must be removed by the blood. In this Investigation, you will design an experiment to test the effects of exercise and other factors on a person's heart rate, which is the number of heartbeats per minute.

PROBLEM

Does a shift in body position cause a person's heart rate to change? How does exercise affect heart rate? After exercising, how long does it take for a person's heart to recover, or return to its resting rate?

HYPOTHESES

Write a **hypothesis** about how exercise and body position affect a person's heart rate.

_____

_____

_____

_____

OBJECTIVES

- Hypothesize about the effects of body position and exercise on heart rate.
- Compare the effects of different body positions and exercises on heart rate.
- After performing different exercises, determine the heart's recovery time.

POSSIBLE MATERIALS

stopwatch or clock with second hand
exercise equipment
proper clothing and shoes for exercise

PLAN THE EXPERIMENT

1. Devise a procedure that tests the effects of body position and various exercises on heart rate, and determines the heart's recovery time after each exercise.

2. Choose body positions and exercises that are safe for you. **CAUTION:** *If you have any medical conditions that may be aggravated by these movements, inform your teacher.* In your procedure, include how long you will maintain each body position and perform each exercise, as well as how many measurements you will make. Think about what your control will be in the experiment.

What Is the Effect of Exercise on Heart Rate?

Lab 37-1

PLAN THE EXPERIMENT cont.

3. Use the stopwatch to take your lab partner's radial pulse.

4. Use Tables 1 and 2 to record your data, or you may wish to make your own data tables.

5. Write your procedure on another sheet of paper or in your notebook. It should include any types of exercise equipment you use.

CHECK THE PLAN

1. Your experiment should include a control.

2. Be sure that the exercises you do are not time-consuming and cause little disruption to other groups in the classroom.

3. Choose exercises that are safe.

4. *Make sure your teacher has approved your experimental plan before you proceed further.*

5. Carry out the experiment.

DATA AND OBSERVATIONS

Table 1

Body Position	Heart Rate (BPM)

Table 2

Effect of Exercise on Heart Rate			
Type of exercise	Duration of exercise (minutes)	Heart rate after exercise (BPM)	Recovery time (seconds)

What Is the Effect of Exercise on Heart Rate?

Lab 37-1

ANALYSIS

1. What was the control in your experiment? What did it show?

_____

_____

_____

2. What body position resulted in the highest heart rate? Explain why.

_____

_____

_____

3. What type of exercise increased your heart rate the most? The least?

_____

_____

4. Explain why the exercises you identified in your answer to question 3 had different effects on heart rate.

_____

_____

_____

5. What is the relationship between type of exercise, heart rate, and recovery time?

_____

_____

_____

6. Two people who have performed the same exercise for the same amount of time may show different recovery times. Why do you think this is so?

_____

_____

_____

7. What factors, other than exercise and body position, may affect heart rate?

_____

_____

_____

_____

What Is the Effect of Exercise on Heart Rate?

CHECKING YOUR HYPOTHESIS

Was your **hypothesis** supported by your data? Why or why not?

_____

_____

FURTHER INVESTIGATIONS

1. Pool class data to compare the average heart rate of males with that of females in your class.

2. Design an experiment that would allow you to observe the effects of different factors on heart rate such as the effects of time of day, type or amount of food consumed, or stress.

When Does a Chicken Embryo Grow the Fastest?

After a sperm has fertilized an egg, the resulting zygote begins to change. The unicellular zygote divides rapidly and develops into a multicellular embryo. Development and growth continue within the egg, until after 21 days, a chick hatches from the egg.

OBJECTIVES

- Hypothesize about the rate of growth of a chicken embryo.
- Observe the development of a chicken embryo over a period of 72 hours.
- Prepare graphs to show the rate of growth and development of a chicken embryo.

MATERIALS

prepared microscope slides
  of a chicken embryo
  24 hours, 48 hours, and
  72 hours after fertilization

microscope
graph paper
  (2 sheets)

PROCEDURE

Part A. Egg at 0 Hours Incubation

1. Make a hypothesis concerning the incubation time period in which the chick will grow the fastest—0 to 24 hours, 24 to 48 hours, or 48 to 72 hours. Write your hypothesis in the space provided.

2. Figure 1 shows a chicken egg just after fertilization. Examine the figure, which shows the membrane-enclosed yolk, composed of protein and fat.

3. The clear fluid surrounding the yolk contains the protein albumin.

4. The two cordlike structures at each end of the yolk are called chalazae and are made of dense albumin.

5. The blastodisc contains a tiny C-shaped structure, the developing embryo. At this point, the length of the embryo is less than 1 mm.

Figure 1

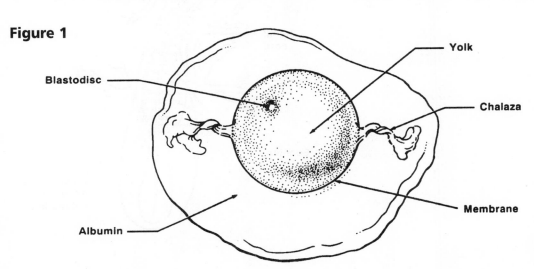

Yolk

Blastodisc

Chalaza

Membrane

Albumin

Embryo after
0 hours incubation

Copyright © Glencoe/McGraw-Hill, a division of The McGraw-Hill Companies, Inc.

When Does a Chicken Embryo Grow the Fastest?

PROCEDURE continued

Part B. Developing Embryos

6. Obtain a prepared slide of a chicken embryo 24 hours after fertilization.

7. Position the slide on the microscope and observe under low power. Use Figure 2 to find the following structures. Locate the neural tube, which runs the length of the embryo. This structure will form the spinal cord. Locate the developing brain, an enlargement at one end of the neural tube.

8. Locate the blocks of tissue forming the neural tube. These are somites. Count and record the number of somites in the embryo and compare it with the number shown in Table 1.

9. Look for the developing heart. It appears as a bulge near the neural tube, between the somites and the brain.

10. Obtain a prepared slide of a chicken embryo 48 hours after fertilization. Observe the slide under low power.

11. Locate the somites and the heart. Count and record the number of somites, and compare it with the number shown in Table 1. As shown in Figure 3, the neural tube has formed the spinal cord and the brain. Locate the eye.

12. Obtain a prepared slide of a chicken embryo 72 hours after fertilization. Observe the slide under low power.

Figure 2

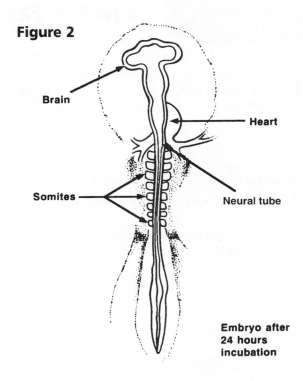

Brain

Heart

Somites

Neural tube

Embryo after 24 hours incubation

Figure 3

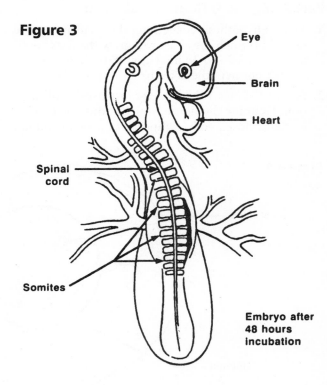

Eye

Brain

Heart

Spinal cord

Somites

Embryo after 48 hours incubation

When Does a Chicken Embryo Grow the Fastest?

13. Find the brain, spinal cord, eye, heart, and somites.

14. Look for paired bulges at the sides of the body, as shown in Figure 4. These bulges are limb buds and will develop into wings and legs. Look for the tail at the end of the body.

15. Prepare a graph that shows the formation of somites during the first 72 hours of development. Plot time in hours on the horizontal axis. Plot number of somites on the vertical axis.

16. Prepare a graph that shows the growth of the embryo during the first 72 hours of development. Plot time in hours on the horizontal axis. Plot length in millimeters on the vertical axis.

Figure 4

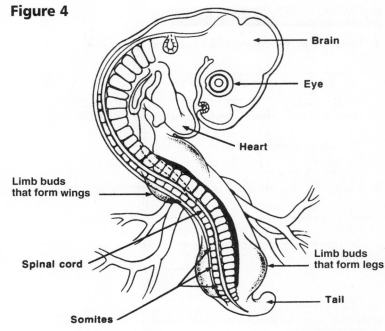

Brain

Eye

Heart

Limb buds that form wings

Spinal cord

Somites

Limb buds that form legs

Tail

Embryo after 72 hours incubation

HYPOTHESIS

_____

_____

_____

_____

DATA AND OBSERVATIONS

Table 1

Hours of Incubation	Length of Chicken Embryo (in mm)	Number of Pairs of Somites	Number of Pairs of Somites Observed
0	Less than 1 mm	0	0
24	4	8	
48	9	21	
72	110	36	

When Does a Chicken Embryo Grow the Fastest?

ANALYSIS

1. What is the function of the yolk and albumin?

_____

_____

2. How does a 72-hour chicken embryo differ from a 24-hour chicken embryo?

_____

_____

3. Using the two graphs you made, determine in which time period the chicken embryo grows the fastest.

_____

_____

_____

CHECKING YOUR HYPOTHESIS

Was your **hypothesis** supported by your data? Why or why not?

_____

_____

FURTHER INVESTIGATIONS

1. Some people are allergic to eggs. Investigate what substances in an egg may cause an allergic reaction. How would having an egg allergy affect a person's diet? Make a list of foods that are safe for someone with an allergy to eggs to consume.

2. Compare the incubation period of a chicken to that of other birds. Is there a correlation between the size of the bird and the length of incubation time?

EXPLORATION

Human Fetal Growth

Complete development of a human fetus takes about 38 weeks. Increases in size and mass are two of the many changes that the fetus undergoes. The increases do not occur at the same rate. Many factors affect the birth size of a human baby, but there is an average mass and an average length standard for each stage of development. The approximate age of a fetus can be determined from its mass and length.

OBJECTIVES

- Calculate the length of a human fetus at various stages of development.
- Graph the length of a developing human fetus.
- Graph the mass of a developing human fetus.
- Determine the period of fetal development during which the greatest changes in mass and in length occur.

MATERIALS

metric ruler

PROCEDURE

Part A. Development of a Human Fetus

1. Examine Figure 1. It shows six stages of a developing human fetus. The stages are shown at 40% of the fetus's actual size.

2. Study the lengths indicated on the diagram of the 38-week fetus. Use these as a guide to measuring the other diagrams.

Figure 1

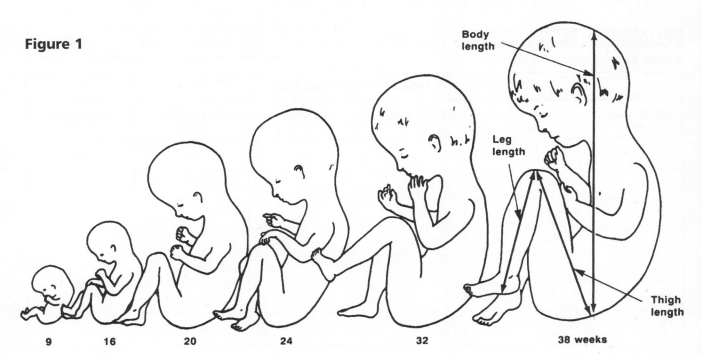

Human Fetal Growth

PROCEDURE continued

3. Measure each length listed below in millimeters. Record your data in the spaces provided in Table 1.

   a. Measure the body length from the rump to the top of the head.

   b. Measure the thigh length from the rump to the knee.

   c. Measure the leg length from the heel to the knee.

4. Add the three measurements for each stage together. Record the total length in the space provided in Table 1.

5. Multiply the total by 2.5 to give a figure that is close to the actual length of the fetus at each stage. Record the actual length in Table 1.

Part B. Graphing the Length of a Developing Fetus

1. The point showing the actual length (2mm) of the 2-week fetus has been marked on the grid in Figure 2.

2. Using the data in Table 1, mark a point that shows the age and actual length of each fetal stage.

3. Begin at 0, and connect the points to complete the graph.

Part C. Graphing the Mass of a Developing Fetus

1. Look at the data supplied in Table 2.

2. Mark points on the grid in Figure 3 to show the age and mass of each fetal stage.

3. Begin at 0, and connect the points to complete the graph.

Table 2

Mass of a Developing Fetus			
Time (weeks)	Mass (grams)	Time (weeks)	Mass (grams)
4	0.5	24	650
8	1	28	1100
12	15	32	1700
16	100	36	2400
20	300	38	3300

DATA AND OBSERVATIONS

Table 1

Length of a Developing Fetus					
Age of fetus in weeks	Body length (mm)	Thigh length (mm)	Leg length (mm)	Total length (mm)	Actual length (mm)
2	—	—	—	—	2
9					
16					
20					
24					
32					
38					

Human Fetal Growth

Figure 2

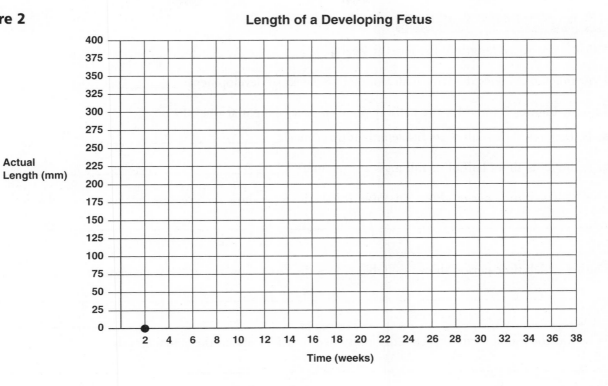

Length of a Developing Fetus

Figure 3

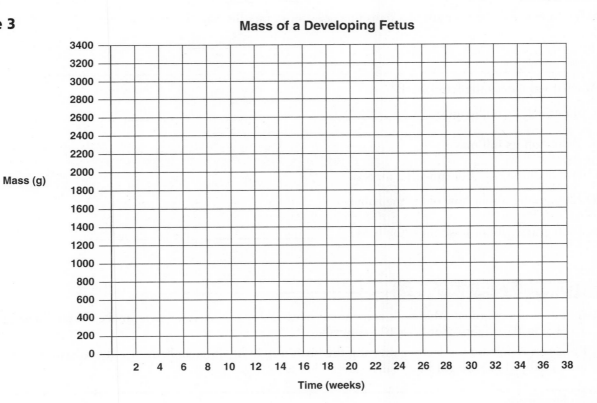

Mass of a Developing Fetus

Human Fetal Growth

ANALYSIS

1. What is the actual length of the fetus at week 9? _____

2. How much mass does the fetus gain from 0 to 8 weeks of development? _____

3. Look at Figures 2 and 3 for the halfway point in development at week 19.

   a. Is the fetus about half of its full length at this time? _____

   b. Is the fetus about half of its full mass at this time? _____

4. During which time period shown in Table 1 does the fetus show the greatest increase in actual length?

_____

_____

5. During which time period shown in Table 2 does the fetus show the greatest increase in mass?

_____

_____

6. Why was the total length of each fetus multiplied by 2.5 to obtain the actual length?

_____

_____

7. Why do you think the length of a fetus increases more rapidly than the mass of a fetus?

_____

_____

8. At what week does the fetus reach

   a. about half its full length? _____

   b. about half its full mass? _____

9. If a premature baby is born with a mass of

   a. 2200 grams, about how old is the baby? _____

   b. 1800 grams, about how old is the baby? _____

FURTHER EXPLORATIONS

1. A human is a mammal. Mammals come in all sizes. Using library references, make a table showing the average birth weights of at least 10 mammals.

2. Prepare a report on the factors affecting the birth weight of a human baby. See if you can find data on how a baby's birth weight might affect his or her later life.

How Do Antimicrobial Substances Affect Bacteria?

In 1928, Sir Alexander Fleming, a British bacteriologist, made a startling discovery. Some mold had fallen accidentally onto one of his bacterial cultures. Fleming noticed that the bacteria did not grow near the mold. He experimented with the mold and isolated it. His experiments revealed that the mold, *Penicillium notatum*, produced a substance that stopped the growth of bacteria. Fleming purified the substance from the mold and developed penicillin, the first antibiotic. Antiseptics, disinfectants, and antibiotics are antimicrobial substances that either inhibit the growth of microbes or kill the organisms directly. Substances that inhibit the growth of bacteria are called bacteriostatic agents. Substances that kill bacteria are referred to as bactericides. The antibiotic penicillin and the disinfectant chlorine bleach are examples of bactericides. Hydrogen peroxide, used as both an antiseptic and as a disinfectant, is bacteriostatic. In this lab you will investigate how several antimicrobial substances affect *Escherichia coli*, the common intestinal bacterium, and *Streptococcus lactis*, the bacterium commonly found in milk.

OBJECTIVES

- Hypothesize how various antimicrobial substances affect the growth of intestinal bacteria and milk bacteria.

- Compare the effects of different antimicrobial

substances on the growth of milk bacteria and intestinal bacteria.

- Use sterile techniques for handling bacterial cultures and inoculating agar plates.

MATERIALS

disinfectant solution
   (25 mL)
paper towels (2)
sterile cotton swabs (4)
forceps
Bunsen burner
striker
sterile filter paper
   disks (14)

test tubes (6)
sterile petri dishes
   containing nutrient
   agar (2)
sterile petri dishes
   containing lactose
   agar (2)
sour milk in a test tube
   (25 mL)

Escherichia coli culture
isopropyl alcohol
   (10 mL)
household bleach
   (10 mL)
tincture of iodine
   (10 mL)
3% hydrogen peroxide
   (10 mL)

antibiotic disks (2)
mouthwash (10 mL)
test-tube rack
masking tape
metric ruler
wax marking pencil
laboratory apron
goggles

PROCEDURE

CAUTION: *Do not touch eyes, mouth, or any other part of your face while doing this lab.*

Part A. Inoculating the Agar Plates

1. Put on a laboratory apron

2. Wash your work surface with disinfectant solution, using a paper towel.

3. Place the four petri dishes containing agar medium in front of you. Label each lactose agar dish with an L and each nutrient agar dish with an N.

Turn the dishes upside down. Be careful not to open the dishes. Use a wax marking pencil to draw perpendicular lines that divide each dish into four equal sections.

4. Number the four sections of the first lactose agar dish 1 through 4, as shown in Figure 1. Number the sections of the second lactose agar dish 5 through 8. Number the sections of the two nutrient agar dishes the same way. Turn the dishes right side up.

How Do Antimicrobial Substances Affect Bacteria?

Lab 39-1

PROCEDURE continued

5. Uncap the test tube of sour milk, flame the mouth of the test tube, and dip a sterile swab into the milk. Continue to hold the swab while you reflame the mouth of the test tube and recap the sour milk. Open one of the lactose agar dishes.

CAUTION: *Always be careful around open flames. Do not let the cotton end of the swab touch the outside of the dish or any other surface. Do not let your fingers touch any sterile surfaces. Use care when working with live bacteria. Avoid spillage on skin, clothing, or work area. Call your teacher immediately if spills occur.*
Move quickly, but carefully, when you inoculate the agar plates and later when you introduce the disks. Lift the lid of the petri dish only when necessary. When introducing the disks or inoculating the plates, hold the lid in one hand and the forceps or swab in the other hand. Tip the lid just enough to permit entry of the swab or forceps. Do not allow the forceps or swab to touch the outside of the dish. Replace the lid immediately.

6. Move the swab lightly back and forth over the entire agar surface in a tight s-shaped pattern. Continue to hold the swab and replace the lid.

7. Turn the petri dish one quarter turn, lift the lid, and repeat step 6, as shown in Figure 2. Dispose of the swab as directed by your teacher.

8. Inoculate the second lactose agar petri dish, following steps 5–7 and using a new sterile swab.

9. Inoculate the nutrient agar petri dishes with *Escherichia coli* using the same procedure as in steps 5–7.

Part B. Testing the Effects of Antimicrobial Substances on Bacteria

1. Obtain the five antimicrobial substances you will test. Place each in a separate test tube and stand the test tubes in a test-tube rack.

2. Sterilize the ends of the forceps by holding the end in the flame of a Bunsen burner for 3 seconds. Use the sterile forceps to pick up a sterile untreated filter paper disk.

3. Tip the lid of the first lactose agar dish and place the disk near the outside edge of section 1. Use the end of the forceps to gently press on the disk so it adheres to the agar, as shown in Figure 3.

Figure 1

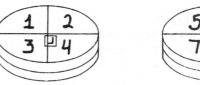
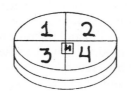
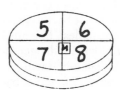

Figure 2

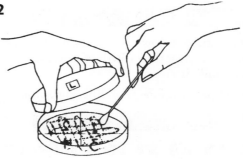

NOTE: *Do not let the forceps touch the agar at any time during this procedure.*

4. Repeat steps 2 and 3, placing the paper disk near the outside edge of section 5.

5. Sterilize the forceps and use them to pick up an antibiotic disk. Place the disk near the outside edge of section 2. Be sure to record which disk you place in which section.

6. Wash the ends of the forceps with water and dry them.

7. Flame the forceps and use them to pick up another sterile filter paper disk. Dip it into isopropyl alcohol and touch the disk to the side of the test tube to drain off the excess alcohol.

8. Place the alcohol disk near the outside edge of section 3 of the lactose agar dish, as in Figure 4.

9. Repeat steps 6–8 with bleach, iodine, hydrogen peroxide, and mouthwash disks in sections 4, 6, 7, and 8, respectively.

10. Repeat steps 2–9 with the nutrient agar dishes.

How Do Antimicrobial Substances Affect Bacteria?

Figure 3

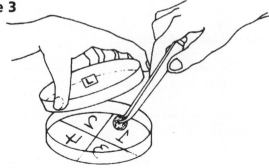

Figure 4

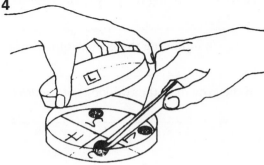

11. Secure the petri dish lids with tape and label them with your name.

12. Make a **hypothesis** to describe how the various antimicrobial substandes will affect the growth of bacteria. Write your hypothesis in the space provided.

Figure 5

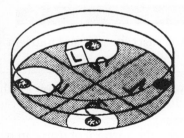

13. Incubate the dishes upside down for 48 hours at 35–37°C.

14. After 48 hours, examine the dishes. DO NOT OPEN THEM. Bacterial growth appears as a cloudy film on top of the agar. Look for zones of inhibition around the disks. These zones are clear areas that indicate an absence of bacterial growth, as shown in Figure 5. Measure the diameter of each zone in millimeters. Record your measurements in Table 1.

15. After completing your observations, dispose of the bacterial cultures as instructed by your teacher. Clean your work surface with disinfectant solution.

HYPOTHESIS

_____

_____

DATA AND OBSERVATIONS

Table 1

Diameter of Zone of Inhibition (mm)		
Disk	*Streptococcus lactis*	*Escherichia coli*
1. Untreated		
2. Antibiotic		
3. Isopropyl alcohol		
4. Bleach		
5. Untreated		
6. Iodine		
7. 3% hydrogen peroxide		
8. Mouthwash		

How Do Antimicrobial Substances Affect Bacteria?

ANALYSIS

1. Why were lactose agar and nutrient agar used to grow the two kinds of bacteria?

2. Why was it important that you lift the lids from the dishes as little as possible?

3. What was the purpose of using an untreated disk of filter paper?

4. Which of the antimicrobial substances were the most effective in inhibiting the growth of *Streptococcus lactis*?

5. Which of the antimicrobial substances were the most effective in inhibiting the growth of *Escherichia coli*?

6. Why do you think the substances had different effects on the growth of the two kinds of bacteria?

7. Why do you think tincture of iodine is used to clean wounds?

8. Did the antibiotic have the same effect on the growth of the two different kinds of bacteria?

9. Based on your observations, do you think an antibiotic, such as tetracycline, can effectively help your immune system fight diseases caused by different types of bacteria? _____

10. Based on your observations, can you conclude which antimicrobial substances would kill pathogenic bacteria?

11. Can all antimicrobial substances be used to kill bacterial diseases in humans? Explain.

CHECKING YOUR HYPOTHESIS

Was your **hypothesis** supported by your data? Why or why not?

FURTHER INVESTIGATIONS

1. Design and conduct an experiment to test the effect of salt on the growth of bacteria. Certain bacteria can cause food to spoil. Based on the results of your experiment, suggest a way to prevent food from spoiling.

2. Design and conduct an experiment to test whether or not an antibiotic continues to have the same effect on the growth of bacteria over time. Explain how you think this information affects what antibiotics a doctor might prescribe for you over time.

Design Your Own

INVESTIGATION

Where Can Microbes Be Found?

Lab
39-2

No matter how hard we try to clean and disinfect areas, microbes return with vigor. Microbes are microorganisms, such as bacteria, molds, and yeasts. More microbes are found in some places than in others. Since infectious diseases are caused by microbes, it's important to know where they are most likely to be found so we can limit their numbers and control the spread of disease. In this Investigation, you will design an experiment to compare the number of microbes found in different locations.

PROBLEM

Where are the most microbes found in public places? Where are the most microbes found in your home?

HYPOTHESES

Write a **hypothesis** indicating in which public and private locations you think the most microbes can be found.

_____

_____

OBJECTIVES

- Make a hypothesis about where the most microbes can be found on surfaces located inside the home and in public areas.
- Collect microbial samples from different locations.

- Distinguish among different kinds of microbial colonies.
- Compare the number of different microbial colonies grown from samples taken from different locations.

POSSIBLE MATERIALS

sterile petri dishes containing
    nutrient agar
sterile cotton swabs
resealable plastic bags
warm location
chlorine solution for cleaning/
    disinfecting work surfaces
paper towels
surfaces of various locations inside and outside
    the home (desktop, drinking-fountain handle,
    computer keyboard, doorknob) from which to
    obtain samples
marker
tape
laboratory apron
goggles

Figure 1

Where Can Microbes Be Found?

PLAN THE EXPERIMENT

1. Choose locations inside and outside of your home to test for the presence of microbes. Be sure to test the locations you listed in your hypothesis.

2. Decide on a procedure to use for collecting microbes and growing them in agar petri dishes. Recall from Investigation 39–1 how to use sterile techniques. If you did not do Investigation 39–1, the teacher will explain sterile techniques to you.

3. Decide how you will record your data and when you will record it. You might use Figure 3 to illustrate the growth of microbial colonies in your petri dishes. Mold, yeast, and bacterial colonies may appear if you allow the petri dishes to sit for several days in a warm location. Bacteria will grow in small circular colonies, whereas molds will spread out more and may look fuzzy. Yeasts tend to grow initially in tight, compact colonies and their color is somewhat darker than that of bacteria. You may wish to summarize your findings in Table 1 or make your own table.

4. Write on another sheet of paper or in your notebook your procedure for collecting and growing microbes. Include the materials you will use.

CHECK THE PLAN

1. Be sure that a control group is included in your experiment and that the experimental groups vary in one way only.

2. Do not open the petri dishes until you are ready to swipe them with the cotton swabs. **CAUTION:** *After swiping the dishes, tape them closed. Do not open them again for the remainder of the experiment. Be sure to wash your hands with soap and water after handling the petri dishes at any time.*

3. *Make sure the teacher has approved your experimental plan before you proceed further.*

4. Carry out the experiment.

5. When you have completed the experiment, dispose of the petri dishes as directed by your teacher.

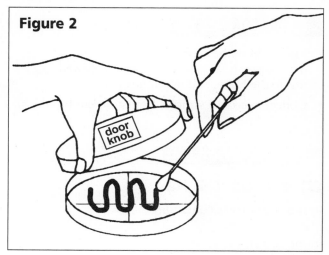

Figure 2

Where Can Microbes Be Found?

Figure 3

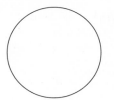
Control

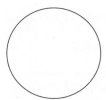
Location:

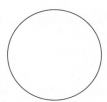
Location:

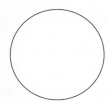
Location:

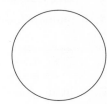
Location:

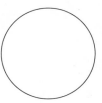
Location:

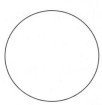
Location:

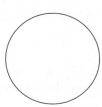
Location:

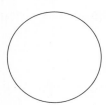
Location:

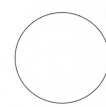
Location:

DATA AND OBSERVATIONS

Table 1

Petri Dish	Bacteria (# of colonies)	Molds (# of colonies)	Yeasts (# of colonies)
Control			
Location:			
Location:			
Location:			
Location:			
Location:			
Location:			
Location:			
Location:			
Location:			

Where Can Microbes Be Found?

ANALYSIS

1. From which sample did you observe the most microbial growth? The least microbial growth?

2. Why do you think there were differences in the number of microbes at the different locations you tested?

3. Describe the control group in your experiment. What did the control group show?

4. What were some possible sources of error in your experiment?

5. Exchange data with another group in your class. What do their data indicate about the presence of microbes inside and outside of the home?

6. How could the information you gained in this Investigation be useful at school or at home?

CHECKING YOUR HYPOTHESIS

Was your **hypothesis** supported by your data? Why or why not?

FURTHER INVESTIGATIONS

1. Disinfect the location you tested that showed the most microbial growth, using a chlorine solution and/or products claiming to be antimicrobial. Then test for the presence of microbes in the disinfected location by repeating the Investigation. Compare your results with the results you obtained previously.

2. Compare the effectiveness of antibacterial soap and regular soap, by collecting samples from your hands.

Glossary

A glossary is the place where you look for the meaning of words. The glossary lists words in alphabetical order. In *Biology: The Dynamics of Life*, *Laboratory Manual*, the glossary has been incorporated to save you time in your search for the meaning of a word. At the same time, the glossary gives you extra reference to related topics and gives you a better opportunity for understanding a word in context. Major vocabulary words used in this manual are listed in the glossary in boldface type, followed by a definition. Some words are followed by a guide to pronunciation in parentheses. Below is a pronunciation key to help you read the words used in this laboratory manual.

Pronunciation Key

a	back (BAK)	**oh**	go (GOH)	**sh**	shelf (SHELF)
ay	day (DAY)	**aw**	soft (SAWFT)	**ch**	nature (NAY chur)
ah	father (FAH thur)	**or**	orbit (OR buht)	**g**	gift (GIHFT)
ow	flower (FLOW ur)	**oy**	coin (COYN)	**j**	gem (JEM)
ar	car (CAR)	**oo**	foot (FOOT)	**ing**	sing (SING)
e	less (LES)	**ew**	food (FEWD)	**zh**	vision (VIH zhun)
ee	leaf (LEEF)	**yoo**	pure (PYOOR)	**k**	cake (KAYK)
ih	trip (TRIHP)	**yew**	few (FYEW)	**s**	seed, cent (SEED, SENT)
i (i + con + e)		**uh**	comma (CAH muh)	**z**	zone, raise (ZOHN, RAYZ)
	idea (i DEE uh)	**u** (+ con)			
			rub (RUB)		

A

abdomen area between the thorax and pelvis in mammals; area behind the thorax in arthropods

abiotic factors nonliving parts of the environment that affect organisms

acid substance that forms hydrogen ions (H^+) in water; an acidic solution has a pH less than 7

adaptation variation in an organism that makes it better able to cope with its environment

albumin protein that makes up egg white

alcoholic fermentation process by which certain yeasts break down sugars in the absence of oxygen

allele (uh LEEL) alternative forms of a gene

allelic frequency percentage of an allele within a gene pool

alternation of generations life cycle in plants, fungi, and some algae in which a haploid gametophyte generation alternates with a diploid generation

ambulacral grooves (am byoo LAK rul) grooves on the ventral side of the rays of a sea star in which the tube feet are found.

ampulla bulblike structure at the base of a tube foot in certain echinoderms

anaphase stage of mitosis and meiosis in which sister chromatids or homologous chromosomes move apart toward opposite poles

antenna sensory organ in many arthropods

anterior toward the front or head end of an animal's body

anther pollen-producing structure of a flower

antheridium male structure that produces sperm

anthophyte plant that produces flowers and forms seeds enclosed in a fruit

antibiotic chemical produced by living organisms that is capable of inhibiting the growth of some bacteria

appendage any major structure growing out of the body of an organism, such as antennae, bristles, and legs

archaebacterium type of prokaryote that lives mainly in harsh habitats

archegonium female structure that produces eggs

arterioles small connecting vessels branching from arteries into capillaries

artery blood vessel that carries blood away from the heart

ascus membranous sac in ascomycote fungi inside which ascospores form during sexual reproduction

atria two anterior chambers of the heart that receive blood from the lungs and body organs

Glossary

B

bactericide substance that inhibits the growth of bacteria

base substance that forms hydroxide ions (OH⁻) in water; a basic solution has a pH greater than 7

basidium microscopic, clublike structure that produces basidiospores in some fungi

bast fibers thick-walled cells in plant stems that surround the phloem and support the stem

Benedict's solution used to test for the presence of simple reducing sugars

binocular vision type of vision that results from both eyes facing forward, which allows for depth perception

biodegradable the ability of a material to be broken down by decomposers

biotic factors living parts of the environment that affect organisms

biuret reagent used to test for the presence of proteins

blood fluid tissue of the body, contained within the circulatory system that carries gases, nutrients, and wastes

brachiopods solitary, bivalve marine animals

brain center of control in the body

bromothymol blue solution used to detect dissolved carbon dioxide

C

caecum also cecum; cavity with only one opening; intestinal pouch in some mammals containing the microorganisms for digestion of cellulose

Calorie amount of heat required to raise the temperature of one milliliter of water one degree Celsius

camouflage protective adaptation that allows an organism to blend with its surroundings

canaliculi small channels in compact bone that carry fluids between the blood vessels and the osteocytes

cancer uncontrolled cell division

canines conical teeth located between the incisors and first bicuspids

capillary thin-walled vessel of the circulatory and lymphatic systems through which gases, nutrients, and wastes are exchanged with body cells

carbohydrate carbon-containing compound, such as sugar, starch, and cellulose

carbonic acid weak acid that forms when carbon dioxide dissolves in water

carrying capacity maximum number of organisms of a species that the environment can support

cell wall rigid structure that surrounds the plasma membrane in certain organisms such as plants, most bacteria, and some protists

centromere point at which sister chromatids are held together

cephalothorax anterior section of some arthropods composed of a fused head and thorax

cerebellum one of three main portions of the brain; controls balance, posture, and coordination

cerebrum one of three main portions of the brain; controls conscious activity, memory, language, and the senses

chalazae two spiral bands of tissue in a bird's egg, connecting the yolk to the lining membrane

chemotaxis characteristic movement of an organism relative to a chemical substance

chlorophyll green pigment that traps light, found in photosynthetic organisms

chloroplast plastid in photosynthetic organisms that contains chlorophyll

chromatography separation of complex mixtures through a selectively absorbing medium

chromatophores pigment-containing structures in the skins of fishes, frogs, and other animals

chromosome DNA-containing structure in a cell nucleus

circulatory system system of structures by which blood is circulated throughout the body

classification systematic arrangement of plants and animals into categories

cloaca in frogs and some other animals, the chamber into which the digestive, excretory, and reproductive systems empty

coelom (SEE lum) true body cavity completely surrounded by tissues

collar encircling structure suggestive of a collar, immediately above the siphon in the mantle of a squid

colon large intestine

colony large group of bacteria descended from a single bacterium

community combination of all the populations of organisms living and interacting with one another in a given area

concentration amount of a substance in a given area or volume

conditioned response learned response to a new stimulus

conditioning process by which an animal learns to respond to a new set of conditions

cones cylindrical structures borne by certain seed plants, consisting of a cluster of stiff, overlapping, woody scales, between which are the naked ovules

constant something that does not change

Glossary

control standard in an experiment by which all conditions are kept the same

conidia asexual spores produced by Ascomycote fungi

consumer organism that obtains energy from eating other organisms

cork outermost layer of cells of a stem that protects against water loss

cork cambium tissue that produces cork on a stem

cortex storage tissue of a stem or root

cusps prominence on the chewing surface of a tooth

D

density mass per unit volume of a substance

deoxyribonucleic acid (DNA) nucleic acid containing deoxyribose as the sugar component and found in the chromosomes of cells

depth of field the depth that one can see clearly under a microscope; foreground or background

development sum of changes that take place during the life on an organism

diaphragm aperture that controls the amount of light entering a microscope

diastase enzyme found in seeds that breaks down starch into maltose

diffusion particle movement from an area of higher concentration to an area of lower concentration

dihybrid cross cross involving two different characteristics

disaccharide condensation product of two monosaccharides bonding together with the elimination of a water molecule

DNA *See* deoxyribonucleic acid

dominant trait trait that is expressed in an organism that it heterozygous for a certain gene

dormant relatively inactive condition in which some processes are slowed down or suspended

dorsal located near or on the back of an animal

duodenum part of the small intestine that extends from the lower end of the stomach

E

ecosystem all the interactions among the populations in a community and the community's physical surroundings

egg female sex cell or gamete

electrophoresis movement of particles in an electric field

embryo early stage of development in any multicellular organism

endodermis innermost layer of the cortex in a root; contains a waxy substance that prevents water from moving out of the central cylinder of a root

enzyme protein that changes the rate of a chemical reaction without being permanently changed or used up by the reaction

epidermis outer, thin protective layer of cells in plants and animals; skin

esophagus tubular portion of the digestive system that connects the mouth to the stomach

eubacteria type of prokaryote that lives in most habitats except harsh ones

eustachian tube canal that connects the pharynx to the middle ear through which air pressure is equalized

external nares external openings in vertebrates for breathing; nostrils

F

fat bodies structures found in frogs on the ovary or testis that store fat as nutrition for developing gametes

fermentation anaerobic process in cells that follows glycolysis and produces ATP

fetus in humans, the term referring to the embryo after the eighth week of development when all systems are present

field of view area that can be seen through a microscope

fossil preserved remains, traces, or imprints of an organism that lived in the past

frond leaf of a fern

fungus unicellular or multicellular eukaryote that has a cell wall and obtains food by absorption

G

gamete sex cell; formed when the number of chromosomes is halved during meiosis

gametophyte haploid form of plants, fungi, and some algae that contains gamete-producing organs

gastrovascular cavity sac in cnidarians in which both digestion and circulation take place

gel electrophoresis method used to separate molecules of different sizes and charges; the molecules are separated by moving through a gel placed in a solution with an electrical field

gene segment of DNA that controls a hereditary trait by coding for a sequence of amino acids in a protein

gene pool all the alleles in a population's genes

genotype genetic makeup of an organism, which determines the traits of the organism

geotaxis movement of an organism in response to the force of gravity

germination development of a seed into a new plant

gill respiratory structure in aquatic animals that removes oxygen from water

glottis opening that leads to the trachea of the respiratory system

Glossary

glucose type of sugar used by cells to obtain energy

gonad organ that produces gametes; testis or ovary

H

Haversian canals small channels containing the blood vessels that nourish the osteocytes in compact bone

Haversian system elongated structure that surrounds a Haversian canal

head upper, anterior vertebrate extremity, containing the brain or principal ganglia

heart muscular organ that pumps blood to all parts of the body

heart rate number of heartbeats per minute

heterozygous condition in which two homologous chromosomes have different alleles for a trait

holdfast structure resembling a root that serves as an anchor for many algae

homozygous condition in which two homologous chromosomes have identical alleles for a trait

hypothesis possible solution to a problem, based on all currently known facts; a prediction that is testable

I

incomplete dominance condition resulting when neither of two alleles is completely dominant; phenotype of the heterozygote is intermediate between those of the homozygotes

inheritance process of genetic transmission of characteristics

ink sac structure in squid that extends from the intestine and releases a dark liquid used for defense

innate behavior behavior that is inherited

intestine portion of the digestive tract extending from the stomach to the rectum

invertebrate animal without a backbone

K

karyotype photograph of paired human chromosomes arranged by size and shape; used to identify chromosomal abnormalities

key table for identifying organisms

kidneys organs that filter waste from the blood

L

lens transparent structure in the eye that helps focus images

ligament fibrous tissue that connects bone to bone

limiting factor environmental factor that affects an organism's ability to survive in its environment

lipid organic compound made by cells for long-term energy storage

liver organ that secretes bile and acts in formation of blood and metabolism; breaks down substances such as alcohol and drugs

lung spongy, saclike organ where breathing occurs

M

macroorganism plant or animal large enough to be seen with the naked eye

madreporite (muh DREH puh rite) delicately perforated sieve plate at the distal end of the stone canal in echinoderms

malaria disease caused by a protozoan that is spread by mosquitoes; symptoms include chills and fever

mantle thin membrane that surrounds the digestive, excretory, and reproductive organs of mollusks and secretes a shell in shelled species

mantle cavity space formed by the mantle that hangs down over the back and sides of the body of the mollusk

maxillary teeth small, conical teeth in the upper jaw of a frog that aid in holding prey

medulla oblongata portion of the brain stem that controls involuntary activities

megaspore reproductive cell in plants that gives rise to the egg

meiosis process of nuclear division that results in the formation of haploid gametes by reducing the number of chromosomes by one half through two divisions of the nucleus

metabolism all the chemical reactions that occur within an organism

metaphase phase in mitosis and meiosis during which sister chromatids or homologous chromosomes line up on the midline of the equator

microbe microorganism, such as a bacterium, mold, or yeast

microscope optical instrument that uses lenses to produce magnified images of small objects

microspore reproductive cell in plants that gives rise to sperm

molars teeth with a broad crown for grinding

monosomy absence of a chromosome

motile having the ability to move

mouth oral cavity; body opening through which an animal takes in food

mutation change in the sequence of a DNA molecule

mycelium mass of hyphae that comprises a fungus growth

Glossary

N

natural selection mechanism for change in populations that occurs when organisms with traits most favorable for survival in a particular environment pass these traits on to their offspring

nematocyst on tentacles in cnidarians, a capsule containing a sharp barb that delivers a poison for obtaining food or for self-defense

nicitating membrane membrane in frogs that keeps the eyeball moist

nosepiece part of a microscope, often rotatable, to which one or more objective lenses are attached

nostril either of two openings of the nose

nucleic acid large, complex macromolecule that stores information in the form of a code; DNA and RNA

nucleotides individual monomers that link together to form a nucleic acid; made up of a nitrogen base, a sugar, and a phosphate group

nutrients carbohydrates, fats, proteins, vitamins, and minerals

O

objective lens system in a microscope that is closest to the object being viewed

osmosis diffusion of water through a selectively permeable membrane

osteocytes bone cells

ovary organ in an animal that produces eggs

P

pancreas gland behind the stomach that secretes digestive enzymes and produces insulin

parasitism symbiotic relationship in which one species benefits at the expense of the other species

pedicellaria (ped uh sehl AH ree uh) small grasping structure on an echinoderm

pen vestigial internal shell of a squid

pH number on a scale that measures the acidity or alkalinity of a solution; a neutral solution has a pH value of 7, an acid solution has a pH value less than 7, and an alkaline (basic) solution has a pH value greater than 7

pharynx upper expanded portion of the digestive tube between the esophagus and the mouth and nasal cavities

phenotype physical appearance of an organism that results from its genotype

phloem (FLOH em) complex food-conducting vascular tissue in higher plants

photosynthesis process by which plants and some bacteria and protists use light energy to make carbohydrates

phototaxis movement of an organism in response to light

pigment molecule that absorbs a specific wavelength of light

plasma membrane outer layer of lipids and proteins that regulates what enters and leaves the cell

plasmolysis shrinking of a cell due to water loss

polytene chromosome giant chromosome found in the salivary glands of fruit flies that is formed by multiple chromosome replications without any separation of the replicated chromosomes

population all organisms of one species in a specific area

posterior back surface of the body; toward the tail

ppm abbreviation for "parts per million"; a unit of concentration

predation capturing of prey as a means of feeding

premolars one of eight bicuspid teeth, behind the canines and in front of the molars

primates mammals such as humans, monkeys, and apes

producer organism that makes its own food; an autotroph

prophase initial stage of mitosis or meiosis in which duplicated chromosomes become visible

protein complex nitrogen-containing organic compound of high molecular weight that has amino acids as its basic structural unit

prothallus heart-shaped gametophyte stage of a fern; develops from a protonema

pulse surge of blood through an artery caused by a heartbeat

Punnett square (PUN ut) tool used to predict the possible offspring of crosses between different genotypes

pyloric sphincter muscular valve that controls the passage of food out of the stomach

pyrenoids small protein bodies that store starch in chloroplasts

R

radial canal in sea stars, one of five pipelike structures that deliver water to the tube feet

radula tonguelike organ in snails with rows of teeth used for scraping food

rays arms of an echinoderm

recessive trait trait that is not expressed in an organism that is heterozygous for a certain gene

rectum last section of the digestive system; in the frog, the rectum opens into the cloaca

reducing sugars monosaccharides and some disaccharides that yield a positive Benedict's test

Glossary

respiration process by which food molecules in a cell are broken down to release energy; also refers to the process of breathing

response reaction of an organism to a stimulus

restriction enzyme bacterial protein that cleaves a DNA molecule at a specific nucleotide sequence

retina photoreceptive layer of cells in the eye

rhizoid anchors nonvascular plants to the soil

rhizome underground stem

root hair single-celled hairlike outgrowth of the root's epidermis; absorbs water nutrients from soil

S

seed plant structure protecting the plant embryo

sessile type of organism that remains permanently attached to a surface for its entire adult life

sickle-cell anemia disease caused by a mutant gene for hemoglobin, which results in sickle-shaped red blood cells that clog small blood vessels and have a shorter life span than normal red blood cells

skeleton system of bones and joints that provides protection and support and allows body movement

small intestine part of the intestine between the outlet of the stomach and the large intestine, where most digestion and nutrient absorption take place

solvent substance that dissolves another substance

sperm male sex cell or gamete

spinal cord bundle of nerves of the central nervous system that extends down the back

sporangium (spuh RAN jee um) cell in which asexual spores are produced

sporophyte spore-producing phase in algae and plant life cycles

stimulus condition that provokes a reaction from an organism

stomach muscular, pouchlike organ of the digestive system

stomata small pores in the surfaces of leaves

suspension mixture in which particles of solid are dispersed in a solid, liquid, or gas

swimmerets appendages on the abdomen of a crustacean

T

telophase: final phase of mitosis and meiosis in which the chromosomes of daughter cells are grouped in new nuclei

tentacles: in cnidarians, the long structures surrounding the mouth used for obtaining food

testcross cross of an individual of unknown genotype with an individual of known genotype, the results from which are used to determine the unknown genotype

testes male reproductive organ of animals

titration process by which the concentration of a substance is determined by comparing it to a standard solution of known concentration

trachea main tubelike structure through which air passes to and from the lungs

transpiration process by which plants lose water through evaporation

trisomy presence of an extra chromosome

tube feet footlike extension of the radial canals controlled by the water vascular system in echinoderms; capable of powerful grasping

tympanic membrane circular structure in a frog's head that responds to air or water vibrations and transmits them to the inner ear and to the brain

U

unicellular organism single-celled organism in which all life functions are carried out

V

variable something that can be changed

vascular tissue tubelike cells in plants through which water, food, and other materials are transported

vein blood vessel that carries blood toward the heart

ventral near the undersurface of an animal body

ventricles two chambers of the heart from which blood circulates to the lungs and body

W

water vascular system in echinoderms, the internal closed system of reservoirs and ducts containing a watery fluid; controls movement, respiration, and food capture

X

xylem principal water-conducting tissue and chief supporting tissue of higher plants

Y

yeast unicellular fungus

Z

zygote diploid cell that is the result of fertilization

Appendix

Figure 1
Pan balances

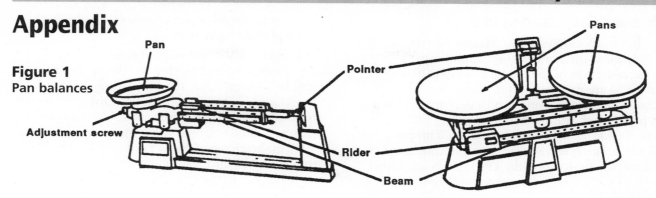

Appendix A. Using the Balance

Although the balance you use may look somewhat different from the balance pictured in Figure 1, all beam balances require similar steps to find an unknown mass.

Follow these steps when using a beam balance.

1. Slide all riders back to the zero point. Check to see that the pointer swings freely along the scale. The beam should swing an equal distance above and below the zero point. Use the adjustment screw to obtain an equal swing of the beams. You should "zero" the balance each time you use it.

2. Never put a hot object directly on the pan. Air currents developing around the hot object may cause massing errors.

3. Never pour chemicals directly on the balance pan. Dry chemicals should be placed on paper or in a glass container. Liquid chemicals should be massed in glass containers.

4. Place the object to be massed on the pan and move the riders along the beams beginning with the largest mass first. If the beams are notched, make sure all riders are in a notch before you take a reading. Remember, the swing should be an equal distance above and below the zero point on the scale.

5. The mass of the object will be the sum of the masses indicated on the beams, as shown in Figures 2 and 3. Subtract the mass of the container from the total mass.

Figure 2
The mass of the object would be read as 47.52 grams.

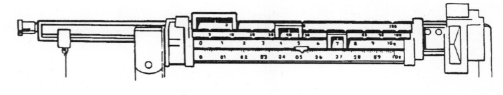

Figure 3
The mass of the object would be read as 100.39 grams.

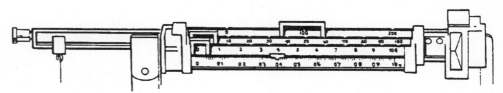

Appendix B. Measuring Volume

The surface of a liquid when viewed in a glass container is curved. This curved surface is called the meniscus. Most of the liquids you will be using form a meniscus that curves down the middle. Read the volume of these liquids from the bottom of the meniscus, as shown in Figure 4. This measurement gives the most precise volume because the liquids tend to creep up the sides of glass containers.

If you are using plastic graduated cylinders and no meniscus is noticeable, read the volume from the level of the liquid.

Figure 4

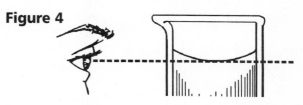

Appendix C. Measuring in SI

The International System of Measurement (SI) is accepted as the standard for measurement throughout most of the world. Four of the base units in SI are the meter, liter, kilogram, and second. The size of a unit can be determined from the prefix used with the base unit name. For example: *kilo* means one thousand; *milli* means one-thousandth, *micro* means one-millionth; and *centi* means one-hundredth. The table below gives the standard symbols for common SI units and some of their equivalents.

Larger and smaller units of measurement in SI are obtained by multiplying or dividing the base unit by some multiple of ten. Multiply to change from larger units to smaller units. Divide to change from smaller units to larger units. For example, to change 1 km to m, multiply 1 km by 1000 to obtain 1000 m. To change 10 g to kg, divide 10 g by 1000 to obtain 0.01 kg.

Common SI Units			
Measurement	**Unit**	**Symbol**	**Equivalents**
Length	1 millimeter	mm	1000 micrometers (μm)
	1 centimeter	cm	10 millimeters (mm)
	1 meter	m	100 centimeters (cm)
	1 kilometer	km	1000 meters (m)
Volume	1 milliliter	mL	1 cubic centimeter (cm^3 or cc)
	1 liter	L	1000 milliliters (mL)
Mass	1 gram	g	1000 milligrams (mg)
	1 kilogram	kg	1000 grams (g)
	1 metric ton	t	1000 kilograms (kg)
Time	1 second	s	
Area	1 square meter	m^2	10 000 square centimeters (cm^2)
	1 square kilometer	km^2	1 000 000 square meters (m^2)
	1 hectare	ha	10 000 square meters (m^2)
Temperature	1 Kelvin	K	1 degree Celsius ($^\circ$C)